РОСТ КРИСТАЛЛОВ

ROST KRISTALLOV

GROWTH OF CRYSTALS

VOLUME 2

Growth of Crystals

Volume 2

A. V. Shubnikov and N. N. Sheftal', Editors

Interim reports between the first (1956) and second Conference on Crystal Growth, Institute of Crystallography, Academy of Sciences, USSR

TRANSLATED FROM RUSSIAN

CONSULTANTS BUREAU, INC. · NEW YORK

CHAPMAN & HALL, LTD. · LONDON

1959

Editors-in-chief

Academician A. V. Shubnikov

♦

Dr. Geol-Min. Sciences N. N. Sheftal'

(Original Russian text published by the Academy of Sciences Press, Moscow, 1959)

ISBN 978-1-4757-0473-0   ISBN 978-1-4757-0471-6 (eBook)
DOI 10.1007/978-1-4757-0471-6

Library of Congress Catalog Card Number: 58-1212

CONTENTS

CONTENTS (continued)

	PAGE	RUSS. PAGE

III. Review and Discussion Articles

FOREWORD

The first volume in the present series contained the papers read at the first conference on crystal growth. This second volume appears during the interval between the first and second conferences, and contains various papers on crystal growth, in addition to the late S. K. Popov's major contribution on growing synthetic corundum.

The papers to some extent represent work that has been done in the USSR since the first conference. Some of the papers break entirely fresh ground.

We hope that the series will act as a means of contact between Soviet scientists who work on crystal growth, or on producing technically important monocrystals.

This contact between scientists is essential. Work on this important topic cannot advance unless such contact is brought about.

A. V. Shubnikov and N. N. Sheftal'

I. THEORETICAL AND EXPERIMENTAL INVESTIGATIONS

THE THERMODYNAMICS OF CRYSTALLIZATION PRESSURE

V. Ya. Khaĭmov-Mal'kov

It has often [1-14] been observed that crystals growing in a medium containing foreign particles repel the particles as well as trap them. Our photographs (Figs. 1, 2, and 3) show such repulsion. Lavalle [4] was the first to notice the effect in 1853. There is as yet no agreement in explanation for the effects, nor is there any agreement in data on the pressures that may occur. Some give values of 10 kg/cm^2 or so [3-5, 10], others only a few g/cm^2 [6, 7, 12, 14]. The latter value has been adopted in monographs on crystal growth [13, 15]. These discrepancies have, of course, meant that fundamentally different explanations can be given for the effect. Those who take the pressures to be small explain the effect as a surface interaction (between crystal, melt, and particle). Those who take the pressures to be large relate the repulsion to the energy of the phase change.

Our object here is to deduce the thermodynamic conditions for a phase change in which a foreign particle is repelled.

Consider the following model. Let the crystal C (Fig. 4) be in a supersaturated solution, and suppose that it has faces a and d covered by a load A, as well as open faces f, c, e, and b.

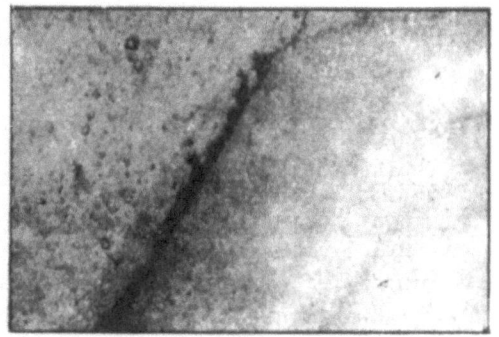

Fig. 1. Face of a thymol crystal growing from the melt and repelling foreign particles. × 12.

The maximum pressure possible at a given supersaturation (or supercooling) is called the crystallization pressure. Numerically, it equals the limiting pressure π the load exerts on 1 cm^2 of the covered face, which pressure just stops growth. The supersaturation (or supercooling) is taken relative to the unloaded crystal. The crystalliza-

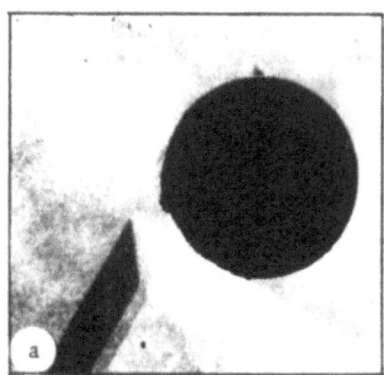 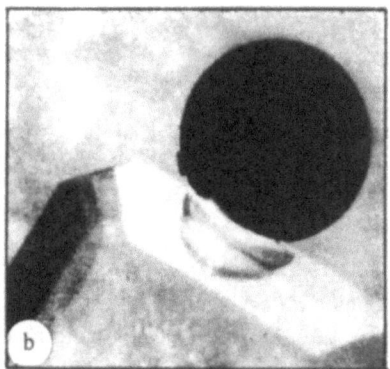

Fig. 2. The (110) face of a potash alum crystal growing from solution and pushing away a mercury drop 0.25 mm in diameter. a) At the start; b) after a certain interval. The liquid inclusion formed under the drop is seen.

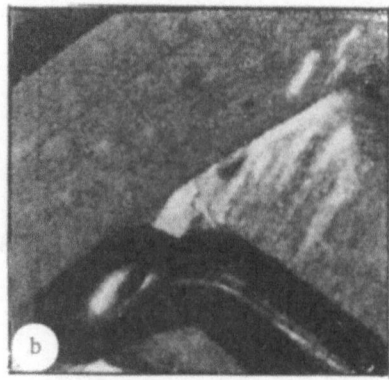

Fig. 3. Crystal of benzophenone growing from the melt and pushing an air bubble to the side. × 12. a) At the start; b) after a certain interval.

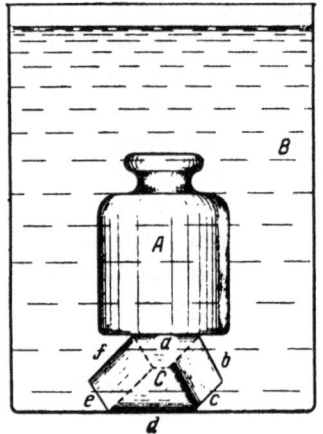

Fig. 4. Model of a phase change in which a foreign particle is repelled.

tion pressure is a one-phase pressure, since it acts only on the obstacles, and not on the obstacle and liquid simultaneously (the limiting pressure acts only on the crystal).

This model can be used to consider the thermodynamics of crystallization pressure in terms of a phase change in which there is an extra one-phase pressure.

1. Phase Equilibrium Conditions for a Crystal Acted on by an Extra Unilateral Pressure

Consider the following simplified model, which we shall use to deduce the phase equilibrium conditions for a crytals acted on by an extra unilateral pressure. Suppose we have a uniform solid body in the form of a cube with its edges parallel to the coordinate axes, which body is in contact with a liquid (vapor) phase only on the faces with z = const and y = const. We assume that the face with z = const is the covered one, and that the one with y = const is the uncovered one. We assume that the solid is uniformly stressed.

We suppose that the system is adiabatically isolated, and that no external work is supplied to it. Then the total entropy S, the total volume V and the total number of molecules N are constant.

These conditions are expressed by the equations

$$
\begin{aligned}
N_1 + N_2 &= N = \text{const}, \\
N_1 v_1 + N_2 v_2 &= V = \text{const}, \\
N_1 s_1 + N_2 s_2 &= S = \text{const},
\end{aligned}
\tag{2}
$$

where v_1 and v_2 are volumes, and s_1 and s_2 are entropies, per molecule of the solid and liquid phases, respectively. The equilibrium condition is

$$
\delta E = 0,
\tag{2}
$$

where δ is the variation operator and E is the total energy of the system. This total energy can be given in terms of the specific phase energies, ϵ_1 and ϵ_2, and of the potential energy E_3 of the load:

$$
E = N_1 \varepsilon_1 + N_2 \varepsilon_2 + E_3.
$$

Hence (2) can be expanded as

$$\sum_{i=1}^{2} (N_l \delta \varepsilon_l + \varepsilon_l \delta N_l) + \delta E_3 = 0 \tag{3}$$

Now δ is a possible change for which (1) is not violated, so the variations δN_l, δv_l, and $\delta s_l (l = 1, 2)$ are related by

$$\sum_{i=1}^{2} \delta N_l = 0,$$

$$\sum_{i=1}^{2} (v_l \delta N_l + N_l \delta v_l) = 0, \tag{4}$$

$$\sum_{i=1}^{2} (s_l \delta N_l + N_l \delta s_l) = 0.$$

We combine (4) with a general definition of the energy [21], and replace $\delta \epsilon_1$ by

$$T_1 \delta s_1 + \sigma_{ik} \delta u_{ik}/N_1,$$

where σ_{ik} and u_{ik} are the stress and deformation tensors, respectively, T_1 is the temperature of the solid, and N_1 is the number of molecules in unit volume of the solid.

Again, we replace $\delta \epsilon_2$ by

$$T_2 \delta s_2 - p_2 \delta u_2,$$

where p_2 is the pressure in the liquid phase, of temperature T_2; and δE_3 we replace by $P\delta h$, where P is the weight of the load, and $\underline{h}$ is the height at which the load lies.

Now $\underline{h}$ equals V_1/f, where f is the area of the face with $z = $ const, and V_1 is the volume of the solid.

But $V_1 = N_1 v_1$, so that

$$h = \frac{N_1 v_1}{f}, \quad \delta h = \frac{N_1 \delta v_1^z + v_1 \delta N_1^z}{f},$$

where δN_1^z is the change in the number of molecules in the solid phase caused by the growth or dissolution along $\underline{z}$ only, and δv_1^z is the change in molecular volume caused by the deformation along the $\underline{z}$ axis only.

The change in E_3 can be put as

$$\delta E_3 = \frac{PN_1 \delta v_1^z + Pv_1 \delta N_1^z}{f}.$$

Now u_{ii} for unit volume is just the change in volume, and δu_{ii} is the element δV_1 of this volume [24], so that

$$\delta v_1 = \frac{\delta V_1}{N_1}.$$

In our case

$$\delta v_1 = \frac{\delta u_{yy} + \delta u_{zz}}{N_1}$$

and so

$$\delta v_1^z = \frac{\delta u_{zz}}{N_1} \, .$$

Now here all the components σ_{ik} are zero, except σ_{yy} and σ_{zz}, so (3) can be put as

$$N_1 T_1 \delta s_1 + \sigma_{zz} \delta u_{zz} + \sigma_{yy} \delta u_{yy} +$$
$$+ \varepsilon_1 \delta N_1 + N_2 T_2 \delta s_2 - N_2 p_2 \delta v_2 - \quad (5)$$
$$+ \varepsilon_2 \delta N_2 + \pi \delta u_{zz} + \pi v_1 \delta N_1^z = 0,$$

where $\pi = P/f$. It is clear that (4) and (5) define the conditions for equilibrium. We use Lagrange multipliers to solve this system. We multiply the equations in (4) in turn by λ_1, λ_2, and λ_3, and combine the resulting forms with (5), grouping the terms appropriately:

$$N_1 (T_1 + \lambda_3) \delta s_1 + N_2 (T_2 + \lambda_3) \delta s_2 + (\sigma_{zz} + \pi + \lambda_2) \delta u_{zz} +$$
$$+ (\sigma_{yy} + \lambda_2) \delta u_{yy} + N_2 (- p_2 + \lambda_2) \delta v_2 +$$
$$\quad (6)$$
$$+ (\varepsilon_1 + \pi v_1 + \lambda_1 + \lambda_2 v_1 + \lambda_3 s_1) \delta N_1^z + (\varepsilon_2 + \lambda_1 +$$
$$+ \lambda_2 v_2 + \lambda_3 s_2) \delta N_2 + (\varepsilon_1 + \lambda_1 + \lambda_2 v_1 + \lambda_3 s_1) \delta N_1^y = 0,$$

where δN_1^y is the change in the number of molecules in the solid phase caused by growth or dissolution only on the face with $y = \mathrm{const}$ (the open face). Now all the variations in (6) are independent, so the equation splits up into

$$T_l + \lambda_3 = 0 \quad (l = 1,2), \quad (7)$$

$$\sigma_{zz} + \pi + \lambda_2 = 0, \quad (8)$$

$$\sigma_{yy} + \lambda_2 = 0, \quad (9)$$

$$- p_2 + \lambda_2 = 0, \quad (10)$$

$$\varepsilon_1 + \pi v_1 + \lambda_1 + \lambda_2 v_1 + \lambda_3 s_1 = 0, \quad (11)$$

$$\varepsilon_1 + \lambda_1 + \lambda_2 v_1 + \lambda_3 s_1 = 0, \quad (12)$$

$$\varepsilon_2 + \lambda_1 + \lambda_2 v_2 + \lambda_3 s_2 = 0. \quad (13)$$

Here (7) implies that $T_1 = T_2 = -\lambda_3$. Also, (10) implies that $\lambda_2 = p_2 = p$; then (8) and (9) give

$$- \sigma_{zz} = p_2 + \pi, \quad - \sigma_{yy} = p_2.$$

Let μ_2 be the chemical potential of the liquid phase; then (13) gives us, when we have substituted for λ_2 and λ_3, that

$$- \lambda_1 \equiv \mu_2 = \varepsilon_2 - T s_2 + p v_2. \quad (14)$$

We substitute for λ_2 and λ_3 in (11) (for the covered face; variation δN_1^z), and in (12) (for the open face; variation δN_1^y), and get

$$\varepsilon_1 - T s_1 + (p + \pi) v_1 = - \lambda_1, \quad (15)$$

$$\varepsilon_1 - T s_1 + p v_1 = - \lambda_1. \quad (16)$$

The conditions for chemical equilibrium on the faces are (15) and (16), which are in general incompatible, because they demand that unequal left halves should equal the same right half (μ_2). Hence, only one of the parallel pairs of faces can be in equilibrium with a liquid of chemical potential μ_2; the other faces cannot be in equilibrium with that liquid.

It is clear that (15) and (16) are not incompatible in the following cases.

1. When π is a hydrostatic pressure. Here the condition for equilibrium is that μ_2 (for the liquid) equals μ_1, the chemical potential of the solid [21].

$$\mu_2(p,\ T) = \mu_1(p + \pi,\ T). \tag{17}$$

2. When the solid is bounded by six surfaces normal to the principal stress axes; surfaces that are not parallel must be in contact with liquids of differing chemical potentials.

3. When the crystal abuts the liquid on one face only. In these latter two cases the normal equality between the chemical potentials of the solid and of the melt is replaced by equality between the η-potential for face $\underline{i}$, and the chemical potential of the liquid that the crystal is in contact with[*]:

$$\eta_{|i} = \mu_2. \tag{18}$$

For example, for our covered face Z

$$\eta_{|z} = \varepsilon_1 - T s_1 + (p + \pi_1) v_1.$$

We now derive expressions for the partial derivatives of the η-potential.

Firstly, (15) implies that, for the covered face,

$$d\eta_{|z} = d\varepsilon_1 - T ds_1 - s_1 dT + (p + \pi) dv_1 + v_1 d(p + \pi). \tag{19}$$

But

$$d\varepsilon_1 = T ds_1 + \mathfrak{s}_{ik} du_{ik}/N_1,$$

or, in our case,

$$d\varepsilon_1 = T ds_1 - \frac{p \, du_{yy}}{N_1} - \frac{(p + \pi) \, du_{zz}}{N_1}, \tag{20}$$

We solve (20) with (19), using the fact that

$$dv_1 = \frac{du_{yy}}{N_1} + \frac{du_{zz}}{N_1},$$

and get for the covered face that

$$d\eta_{|z} = -s_1 dT + v_1 d(p + \pi) + \frac{\pi}{N_1} du_{yy}.$$

We transfer to the usual independent variable for the chemical potential (for which we use Hooke's law $u_{zz} = \frac{\pi}{E}$, and $|u_{yy}| = |\gamma u_{zz}|$, where E is Young's modulus and γ is Poisson's ratio), and we get, for an isotropic solid, that

_____

[*] This name "η-potential" is the one we shall use here.

$$d\eta_z = -s_1 dT + v_1 dp + \left(v_1 + \frac{2\pi\gamma}{EN_1}\right) d\pi. \tag{21}$$

It is clear from (21) that

$$\frac{\partial \eta_z}{\partial T} = -s_1, \quad \frac{\partial \eta_z}{\partial \pi} = v_1 + \frac{2\gamma\pi}{EN_1}, \quad \frac{\partial \eta}{\partial p} = v_1. \tag{22}$$

A like argument for the open faces shows that

$$d\eta_{iy} = -s_1 dT + v_1 dp + \frac{2\pi\gamma}{EN_1} d\pi,$$

whence

$$\frac{\partial \eta_y}{\partial T} = -s_1, \quad \frac{\partial \eta_y}{\partial p} = v_1, \quad \frac{\partial \eta_y}{\partial \pi} = \frac{2\pi\gamma}{EN_1}. \tag{23}$$

We consider the conditions for equilibrium in one- and two-component systems in more detail below, where we use these results.

2. Conditions for Equilibrium in One-Component Systems

We shall consider the conditions for equilibrium in one-component systems for covered and open faces separately. As before, we assume that the applied stresses and the volume are such that (22) and (23) are complied with.

Covered Faces

Now (19), for the η-potential, and for the chemical potential of the liquid, gives us that

$$\frac{\partial \eta}{\partial T} dT + \frac{\partial \eta}{N_1 \partial \sigma_{ik}} d\sigma_{ik} = \frac{\partial \mu}{\partial T} dT + \frac{\partial \mu}{\partial p} dp. \tag{24}$$

We now use (22), and also the fact that

$$\frac{\partial \mu}{\partial T} = -s_2 \quad \text{and} \quad \frac{\partial \mu}{\partial p} = v_2,$$

to get from (24) that

$$Q_1 \frac{\partial T}{T} = (v_2 - v_1) dp - \left(\frac{2\pi\gamma}{EN_1} + v_1\right) d\pi \ *, \tag{25}$$

where Q_1 is the latent heat of the phase change. When $\pi = 0$, we have that (25) becomes simply the Clausius-Clapeyron equation, which defines how the transition temperature varies with the total pressure.

If p = const, (25) implies the conditions for the transision on the covered face when a unilateral pressure is acting. We multiply both parts of (25) by N_1, and so get the same equation, but now referred to unit volume of the solid phase, V_1:

$$Q \frac{\partial T}{T} = (V_2 - V_1) dp - \left(\frac{2\pi\gamma}{E} + V_1\right) d\pi, \tag{25a}$$

---

* It has several times been shown [18, 19, 22] that Hooke's law can be used near the melting point; Young's modulus and critical shearing stress measurements have been used to show this.

where

$$V_2 = N_1 v_2, \quad Q = N_1 Q_1, \quad V = N_1 v_1.$$

We assume that the coefficient to $d\pi$ is independent of π and of temperature; (25a) then gives (with p = const) that

$$\ln \frac{T}{T_0} = -\frac{1}{Q}\left(V_1 \pi + \pi^2 \frac{\gamma}{E}\right), \tag{26}$$

where $T = T_0$ when $\pi = 0$, or, approximately,

$$\Delta T = -\frac{T_0 V_1}{Q} \pi - \frac{T_0 \gamma}{QE} \pi^2. \tag{27}$$

The values of the terms in (27) are such that, if π lies in the range 10-100 kg/cm^2,

$$\frac{T_0 V}{Q} \pi \gg \frac{T_0 \gamma}{QE} \pi^2 \, (\gamma \approx 0.2, \ E \approx 10^5 \ \text{kg/cm}^2),$$

it needs pressures of geological order to cause the second term in (27) to affect ΔT appreciably. Hence, the anisotropy affects only terms of the second order of magnitude, since it enters only via γ and E.

Hence, the change in transition temperature caused by unilateral pressures in the 0-1000 kg/cm^2 range on a closed face is independent of the face.

It can happen that ΔT varies very much with π. Thus, for NaCl (whose melting point at $\pi = 0$ is 804°C, with $v \approx 0.4$ cm^3 and Q = 123.5 cal/g) we find that ΔT is 0.75°C for a rise in π of 10 kg/cm^2.

Now if T = const, (25a) gives us, in the light of the above discussion about (27), that

$$\frac{dp}{d\pi} = \frac{V_1}{V_2 - V_1},$$
$$p = p_0 + \frac{V_1 \pi}{V_2 - V_1}, \tag{28}$$

where p = p_0 when $\pi = 0$.

Equation (28) defines the change in the total pressure for an isothermal phase transition when the unilateral pressure changes; e. g., for mercury, a rise in π from 0 to 1 kg/cm^2 implies a rise in total pressure from 1 to 20 kg/cm^2 (the molar volume of solid mercury at −38.9°C is 14.15 cm^3, and of liquid mercury at 0°C is 14.7 cm^3).

A good example of a single-phase pressure (but an isotropic one) is a surface pressure $2\sigma/r$, where σ is the surface tension of a drop of radius $\underline{r}$. We substitute this π into (26) and get Thomson's formula

$$T = T_0 \exp - (2\sigma v/rQ)$$

or the formula for the critical radius of a drop, r_c:

$$r_c = 2\sigma v T_0/Q\Delta T.$$

The conditions for thermodynamic equilibrium on covered faces thus lead to the following conclusions.

1. The crystallization pressure increases with the supercooling; the covered face of a growing crystal can repel loads that increase with the supercooling.

2. The crystallization pressure is almost independent of the face. Hence, faces subject to differing single-phase pressures cannot all be in equilibrium with a given liquid or vapor at once.

Open Faces

Before we deal with open faces, we note that W. Thomson first formulated a similar problem for the changes in thermodynamic equilibrium parameters caused by stresses in a solid phase. J. Thomson solved this problem for the ice—water system in 1849 [16]. Gibbs gave this solution in a general form in 1879, and Riecke gave it in 1894 [20]. One of the deductions from this solution, namely (30), is called the Gibbs-Riecke formula.

Equations (15) and (16) for the η-potentials show that a covered face has a larger η-potential than an open face does. The open faces, or open parts of a loaded face, can grow when the covered part is in equilibrium with the liquid. We use (18) and (20) to find how the melting point for an open face varies with π; by analogy with (25a), we find that

$$Q \frac{dT}{T} - (V_2 - V_1)\, dp = -\frac{\pi\gamma}{E}\, d\pi. \tag{29}$$

Now (29), with p = const, gives us that

$$T = T_0 \exp -\frac{\pi^2\gamma}{QE},$$

where $T = T_0$ when $\pi = 0$. Or, approximately,

$$\Delta T = -\frac{T_0\gamma}{QE}\, \pi^2, \tag{30}$$

where the $T_0\gamma\pi^2/QE$ term is analogous to the second term in (27).

We find that for NaCl (crystal and melt) $E_{[100]} = 15.6 \cdot 10^4$ kg/cm^2 (E measured at 804°C), so for $\pi = 10$ kg/cm^2 we have $\Delta T = 0.3 \cdot 10^{-4}\, \gamma$. The same π gives $\Delta T \approx 0.7$°C for a covered face.

At T = const, $\underline{p}$ is related to π by

$$\Delta p = \frac{\gamma}{E(V_2 - V_1)}\, \pi^2$$

so, if we take ΔV as about 0.05 cm^3/g, and E $\approx 10^4$–10^6 kg/cm^2, we have, for $\pi \approx$ 10-100 kg/cm^2, that $\underline{p}$ will vary from 10^{-3} to 10 kg/cm^2.

Hence, an elastically strained crystal has a lower melting point than an unstrained crystal, other things being equal; the change in the melting point is independent of the sign of the deformation; and the melting point on a covered face is less than on an open face.

3. Conditions for Phase Equilibrium in Two-Component Systems

The difference from a one-component system is only that an extra thermodynamic parameter C (the concentration) enters into the conditions.

Hence, we use the results in Sections 1 and 2 above, and find, by analogy with (25), that for a covered face

$$Q_1 \frac{dT}{T} - (v_2 - v_1)\, dp + v_1 d\pi = \frac{\partial\mu}{\partial C}\, dC - \frac{2\pi\gamma}{EN_1}\, d\pi.$$

Now the terms in (27) are such that, when $\underline{p}$ and T are constant,

$$\frac{\partial C}{\partial \pi} = v_1 \left/ \frac{\partial\mu}{\partial C} \right. . \tag{31}$$

10

Thus (31) gives us how the equilibrium concentration on a covered face varies with the crystallization pressure during isothermal isobaric crystallization. The solution depends on how μ is related to C.

Dilute Solutions

Now $\mu = kT \ln C - \psi$ for dilute solutions, where ψ is a function that does not depend on concentration, and $\underline{k}$ is Boltzmann's constant. We substitute for μ in (31) and get*

$$C = C_0 \exp \frac{\pi v_1}{kT}, \tag{32}$$

where $C = C_0$ when $\pi = 0$.

Correns was the first to relate C to π, by equating the chemical potentials of the substance in the solid and dissolved states [14]. We have shown above that this treatment is correct only if a single-phase pressure acts all over the crystal. The chemical potentials cannot be equal if such a pressure is applied to one face only. The formulas are of the same form because in (31) we neglected the term $\frac{2\pi \gamma}{EN_1} d\pi$, which accounts for the anisotropy in the applied pressure; we did this because the term was small relative to $v_1 d\pi$.

The conditions for thermodynamic equilibrium thus imply the following conclusions for a covered face in a dilute solution.

1. The crystallization pressure increases with the supersaturation.

2. The crystallization pressure depends little on the properties of the substance (the specific volume is the only parameter that appears).

3. The crystallization pressure is independent of the solvent if the solution is dilute.

4. The crystallization pressure is the same for all faces. Hence, phases subject to differing single-phase pressures cannot be in equilibrium with the same liquid or vapor phase.

Real Solutions

We can use Lewis' method to deal with real solutions; we insert in the chemical potential a function that describes the deviations from the laws for ideal systems. The activity $\underline{a}$ is used as this function in the theory of solutions; it is given by

$$u = \mu_0(T, \; p) + kT \ln a. \tag{33}$$

Now (31) and (33) give us that

$$\pi = \frac{kT}{v} \ln \frac{a}{a_0},$$

where a_0 is the activity at $\pi = 0$.

We can use an empirical formula relating $\underline{a}$ to C in order to compare ideal and real systems. Hildebrand and Eastman deduced this formula in the form [23]

$$\lg \frac{a}{N_1} = \frac{2 \cdot 0.096}{0.263} \left[\frac{1}{1 + 0.263 \, M_2/M_1} - \frac{1}{2(1 + 0.263 \, M_2/M_1)^2} \right],$$

where $\underline{a}$ is the activity of the dissolved substance, and M_1 and M_2 are the molar fractions of the solute and solvent in the solution. The table gives data for $CuSO_4$ in water at 0°C, for real and ideal solutions.

---

* If we replace π by the pressure caused by surface tension, $2\sigma/r$, as before, we get Thomson's formula for solutions.

C/C_0		1.01	1.02	1.1	1.3	1.5	2
π, kg/cm^2	ideal solution	5.05	10.1	47.6	133	206	350
	real solution	6.95	13.9	65.5	182	281	480

It is clear that the real solution has a higher crystallization pressure than does the ideal one, and one that rises more rapidly with concentration, at that; the conditions for thermodynamic equilibrium are otherwise the same.

SUMMARY

A thermodynamic consideration of crystallization pressure shows why the faces of a growing crystal repel foreign particles. However, Correns' data show that the theoretical and actual ways in which π varies with C are not the same. The agreement is good at small and medium supersaturations ($C \le 1.2$), but the theoretical formula gives results that are too high at high supersaturations. Correns' data also indicate that the pressure begins to depend on the face at the latter level, which also contradicts the theory. These features of Correns' data, which do not agree with data given by others, mean that no final answer can be found to the problem until the reasons for the discrepancies have been elucidated. Our experiments* indicate that the higher pressures must be taken as the correct ones.

Our theory is in essence related to Gibbs' deductions about phase equilibrium for deformed bodies [17]; an experimental confirmation of the one is at once a confirmation of the other. The main point is that, if a homogeneous strained body is bounded by six faces normal to the principal axes of the stress ellipsoid, the conditions for mechanical and chemical equilibrium are satisfied only by contact with liquids that exert different pressures on opposed pairs of faces. But whereas it was difficult to test the theory for the cases Gibbs considered (because the effects were small; see Section 2), it has been possible to perform such tests for single-phase pressures on crystals.

It is clear that an extra thermodynamic parameter (the crystallization pressure) should be considered, together with the temperature, concentration and so on, in choosing conditions for phase equilibrium. Thus, for Li (melting point 178.4°C at p = 1.03 kg/cm^2) a single-phase pressure of 100 kg/cm^2 will reduce the melting point by about 60°C (latent heat of fusion 32.8 cal/g), whereas about 8000 kg/cm^2 is needed as a hydrostatic pressure in order to raise the melting point by 24°C.

I should like to thank Academician Shubnikov and Dr. Sheftal' for their help with this work, and also to thank A. A. Chernov, of the Institute of Crystallography, for valuable discussions of the results.

LITERATURE CITED

[1] H. Kopp, Ann. Chem. Pharm. 44, 124 (1855).

[2] K. Andree, Geolog. Rundschau 3, 7-15 (1912).

[3] C. Becker and A. Day, Zentralbl. Miner. 337-346, 364-373 (1916).

[4] J. Lavalle, Compt. rend. 34, 493 (1853).

[5] G. Correns, Sitz. Ber. Preuss. Akad. Wiss. 11 (1926).

[6] A. Shubnikov, Z. Krist. 88, 466-469 (1934).

[7] A. V. Shubnikov, Trudy Lomonosov Inst. Akad. Nauk SSSR No. 6, 17-21 (1935).

[8] C. Benedicks and H. Löfquist, Nonmetallic Inclusions in Iron and Steel [Russian translation] (ONTI, 1935).

[9] A. A. Bochvar, The Mechanism and Kinetics of Crystallization in Eutectic-Type Alloys [in Russian] (ONTI, 1935).

[10] E. Scheil, Z. Metallkunde 27, 4, 76 (1935).

* See the articles that follow this one.

[11] C. Correns and W. Steinborn, Z. Krist. A 101, 117-133 (1939).

[12] I. N. Fridlyander and N. A. Vyostskaya, Doklady Akad. Nauk SSSR 62, 1, 71-3 (1948).

[13] V. D. Kuznetsov, Crystals and Crystallization [in Russian] (Moscow, 1954).

[14] G. B. Bokii, Trudy Inst. Kristall. Akad. Nauk SSSR No. 5, 143-8 (1949).

[15] H. E. Buckley, Crystal Growth [Russian translation] (IL, 1954).

[16] J. Thomson, Trans. Roy. Soc. Edinburgh 16, 5, 579-580 (1848-1849).

[17] J. W. Gibbs, Thermodynamic Works [Russian translation] (IL, 1950), p. 247.

[18] L. Hunter, Phys. Rev. 61, 84-90 (1942).

[19] A. V. Stepanov and N. Eidus, J. Exptl.-Theoret. Phys. (USSR) 29, 669-675 (1955).

[20] E. Riecke, Ann. Phys. und Chemie 34, 731-738 (1895).

[21] P. Niggli, Z. anorg. Chemie 107-133 (1915).

[22] E. Schmid and W. Boas, Plasticity of Crystals, Especially Metal Ones [Russian translation] (ONTI, 1938), p. 134.

[23] C. Lewis and M. Randall, Thermodynamics and the Free Energy of Chemical Substances (1923), p. 271.

[24] L. D. Landau and E. M. Lifshits, The Mechanics of Continuous Media [in Russian] (Gostekhizdat, 1953).

EXPERIMENTAL MEASUREMENT OF CRYSTALLIZATION PRESSURE*

V. Ya. Khaimov-Mal'kov

In the preceding paper** we pointed out that the measured values of crystallization pressure vary greatly. Our tentative theoretical study of growth in a crystal subject to a unilateral pressure has shown that the pressure sufficient to stop growth agrees well with the larger crystallization pressures found by experiment [1].

The way the pressure depends on some thermodynamic parameters also agrees well with experiment. But, unless we can explain why the experimental values disagree so greatly, we shall not have given a final explanation of how a crystal repels an obstacle.

Our object here is to find out why earlier experimental measurements of crystallization pressures gave discordant results.

A Brief Historical Review

Becker and Day [2] were the first to measure crystallization pressures at all exactly, in 1905. They placed a loaded glass plate on top of a growing alum crystal (Fig. 1). The crystal went on growing even when the load

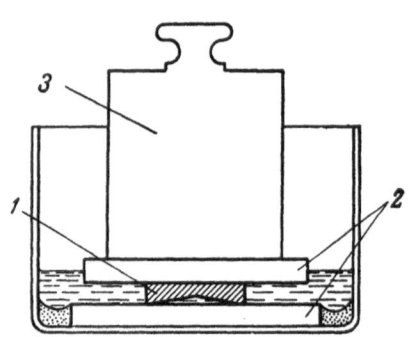

was 1 kg. They described the effect in terms of a "linear crystallization force," and instanced the way pyrite crystals growing in graphite split the latter as being a similar effect. They found similar effects with potassium ferrocyanide, copper sulfate, and lead nitrate.

Later Andree [3] confirmed Becker and Day's idea, and pointed out that completely transparent gypsum crystals could grow in clay masses; this was only one among many geological examples.

In 1913, Bruhns and Mecklenburg [4] claimed that they had repeated Becker and Day's experiments carefully, and that crystallization pressure did not exist. The growth cones formed under the lower face of their crystal (Fig. 2) in the system they used (Fig. 1) they assigned to dissolution, and not to any possible growth at the lower face, which might lift the load. The dis-

Fig. 1. The method used in Becker and Day's experiments. 1) Growing alum crystal; 2) glass plate; 3) load.

solution they considered to occur because there was a layer of solution under the crystal.

In 1914 Taber [5] described his experiments, which he claimed showed that Bruhns and Mecklenburg were wrong, and that Becker and Day were right.

In 1926, Correns [6] confirmed Becker and Day's quantitative results. He found a pressure of 50 g/cm² for the (111) face of alum.

In 1934, Shubnikov [7, see also 8] gave his measured value for alum as 0.41 g/cm². The measurements were made with the growing crystal encountering a glass ball on the end of a thin glass fiber (Fig. 3); the fiber

---

* Academician A. V. Shubnikov proposed this work.

** See the paper in this volume, p. 3.

Fig. 2. Growth funnel on the lower face of a crystal. × 6.

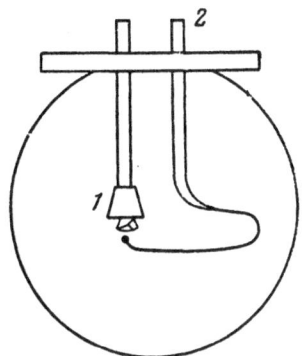

Fig. 3. Scheme of Shubnikov's experiment: 1) bung with crystal mounted on it, 2) glass rod with its end drawn out and ending in a ball.

acted as a spring. The crystallization force was taken to be that force which corresponded to the deflection at which the ball began to enter the crystal. The area of contact (which is needed to calculate the pressure) was taken as the area of the spherical segment, i. e., the area of the visible contact at the end of the experiment. The tests were done at room temperature, with a supersaturation of 2 g per 100 cm^3 of solution.

Shubnikov related the mechanism to interactions at the surfaces where the crystal, ball and solution met (as did also Correns).

Shortal afterwards, Scheil [9] published his work. He pointed out that he had seen Correns' and Shubnikov's work anly after he had finished his own, and gave a method of measuring the crystallization pressure of Zn (and measured values). A piece of steel (cold) lowered into fused Zn at 450-480°C became coated with solid zinc on the faces, whereas the edges stayed almost free. This effect was used in the following way. Two steel wedges were lowered into fused Zn and were coupled to a balance.

The wedges were set about 1 mm apart, and the melt entered this gap freely. The solid Zn then began to build up between the wedges; the resulting force was measured by the balance. Scheil gives graphs of growth rate against load for the range 450-470°C. The growth stopped at 5 kg/cm^2. This can be taken as the crystallization pressure. Scheil took the area as that of the solidified material formed in the gap.

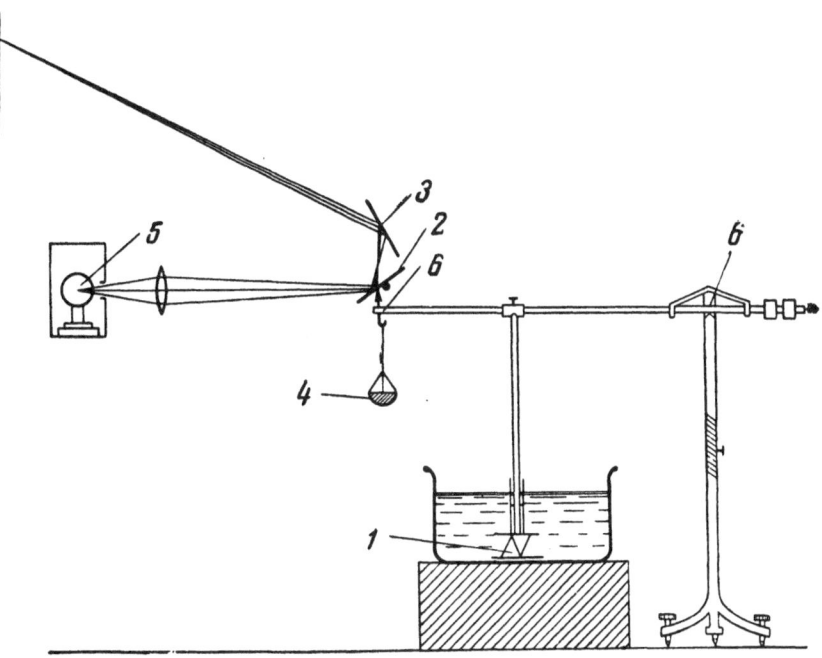

Fig. 4. Correns' and Steinborn's system: 1) crystal; 2, 3) mirrors; 4) load; 5) light source; 6) balance arm.

15

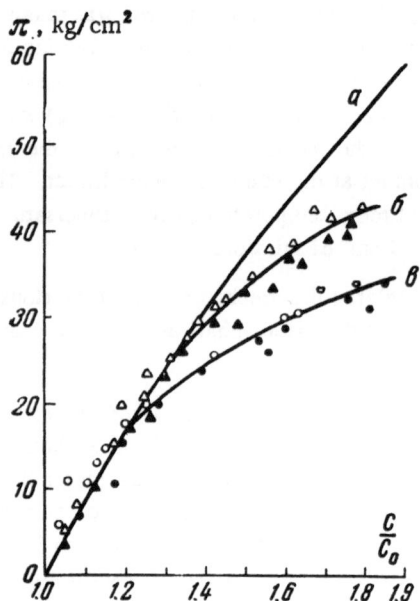

Fig. 5. Relation of crystallization pressure to supersaturation: a) theoretical; b) experimental for (111); c) experimental for (110). Black symbols — crystal growing; open symbols — growth stopped.

In 1939, Correns and Steinborn published another paper on a method of measuring crystallization pressures [1]. They used Becker and Day's method, somewhat improved by a sensitive balance (Fig. 4). They do not give the area of contact, nor the method they used to measure the area.

Figure 5 shows their results for the (111) and (110) faces of alum, as functions of supersaturation. It is clear that the pressure depends on the supersaturation and can rise to 40-50 kg/cm^2, that the pressure depends on the face, and that the experimental curves agree well with the theoretical relation

$$\pi = kT \ln \frac{C}{C_0} ,$$

at small supersaturations (here $\underline{k}$ is Boltzmann's constant, T is the temperature, and C/C_0 is the supersaturation). It was also found that the pressure depended on the obstacle. Thus mica (cleavage plane) gave zero pressure for the (111), (110) and (100) faces at all supersaturations.

In 1948, Fridlyander and Vysotskaya [10] used Shubnikov's method, and found values of about 0.072-0.152 g/cm^2 for salol.

In 1949 Bokii [11] found a value of 0.13-0.17 g/cm^2 for alum. He used a glass plate (instead of the glass ball of Shubnikov's method), so set that the crystal pushed the plate upwards. The area of contact was measured as the width of the last step on the growth funnel at the top. The concentration was such that the solution was saturated at 26-27°C; the crystal grew at room temperature over about three days.

In 1951 Vadilo [12] used a method similar to Becker and Day's to measure the pressure between a (100) face of alum and a glass obstacle. The value he found was 80 kg/cm^2, whereas Correns and Steinborn, and Bokii, had found that the pressure was zero at all supersaturations for this face and a glass obstacle. Correns had also shown that the pressure increased with the supersaturation, whereas the converse is asserted in [12].

A Critical Review of the Methods Used to Measure Crystallization Pressure

The above results for the pressure may be divided into two groups. The results differ by a large factor, and in the ways in which they were obtained. The first group contains results obtained by Becker's method [2, 6, 9, 10] and the second those obtained by Shubnikov's [7, 11].

The difference in the methods is that a glass plate is used in the first, and a ball in the second; in the first, the plate touches a defined face, whereas the contact in the second is not along any crystallographic surface; in the first, the plate is of such a size that the crystal cannot grow past it, whereas in the second, the crystal can bypass the ball; and, finally, in the first, the area of contact is the area of the last step on the growth funnel, whereas in the second, the area used is that of the contact at the end of the experiment.

In view of this, we shall not deal with all experiments, but only with Correns' and Shubnikov's work, as representing the two different methods. This excludes Bokii's work, in which Becker's method was used, but which gave pressures similar to those Shubnikov found (about 0.15 g/cm^2); we shall therefore deal seaprately with this work.

Correns and Steinborn. We repeated these experiments almost in their entirety. In some cases we used exactly the same system as Correns did; in others we used photomultipliers and other methods of recording displacements.

Our results, especially at low supersaturations, were very like Correns'and Steinborn's. At high supersaturations (> 1.25) we found some difficulties that Correns and Steinborn do not mention.

Fig. 6. Glass ball in the growth pyramid of a [111] face on an alum crystal. The position of the ball shows that the ball was at first pushed away. The fiber that holds the ball suspended is seen. × 4.

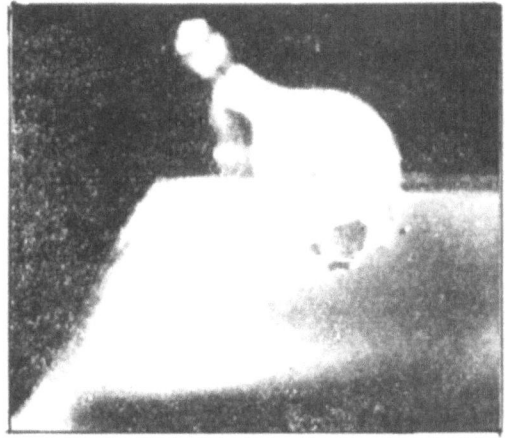

Fig. 7. Repulsion of a glass ball by a (100) face of an alum crystal. A negative crystal is visible under the ball. × 10.

Fig. 8. Pit from a ball which has entered a crystal. × 20.

Firstly, it was very difficult to establish the end point, i. e., the load at which the crystal stopped growing, because the growth rate on the covered face became so low near the endpoint. Hence, the difference between the loads at which the crystal clearly was and was not growing became large (sometimes as much as 30-50%).

The resulting uncertainty in the time of the endpoint implied a corresponding uncertainty in the area of the last step on the growth funnel (Fig. 2), because the width and perimeter depend on the growth of the open faces, which grow rapidly. Our results confirmed Correns' and Steinborn's as to order of magnitude.

Shubnikov. We repeated Shubnikov's experiments many times under steady and changing growth conditions, but never got reproducible results. The following problems arose in this connection.

It is not certain that the force is measured unambiguously, i. e., that the crystallization force (in the sense adopted above) is the force measured in terms of the maximum deflection. It is also not certain that the area to be used is the area of the segment on the sphere.

To solve the first problem, we did the following test. The ball was simply hung on the glass fiber, instead of being fastened on. If the ball is pushed away at first, it should continue to be pushed away, because the applied force (the apparent weight of the ball in the solution) does not change. The pressure on the face also stays the same, or becomes less, because the area of contact may increase. We found, though, that such balls always entered the crystals in time (Figs. 6 and 7). Hence, the crystallization force is not the force applied at the time the ball starts to enter the crystal. The results show that the part of the face under an obstacle grows less rapidly than does an unobstructed area.

The second problem is solved by the observation that the contact surface between ball and crystal is not smooth (which is contrary to the assumption on which the method is based), but resembles a growth funnel. The surface consists of hollows and steps, as the profile (Fig. 9) of a pit (Fig. 8) shows. Hence, we may say that the crystal pushes away the ball not over the entire contact surface, but only over the steps. This effect must be allowed for in measuring the area, because it reduces the area greatly; otherwise the error in the pressure may be two or three powers of ten.

In Correns' experiments, the pressure depended on the face, but no such effect is noticed in experiments of Shubnikov type. We grew two crystals in the same crystallizer, which crystals were placed symmetrically about a slowly rotating (5 rpm) stirrer. The balls nevertheless rose differently. The ball on an octahedron face (Fig. 6)

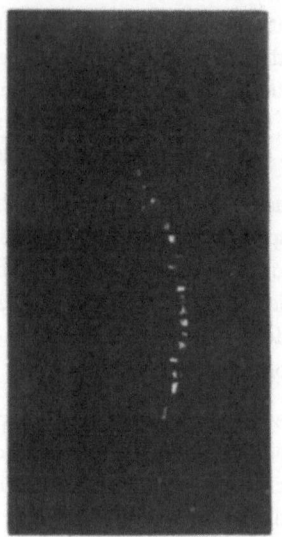

Fig. 9. Profile of a pit. × 100.

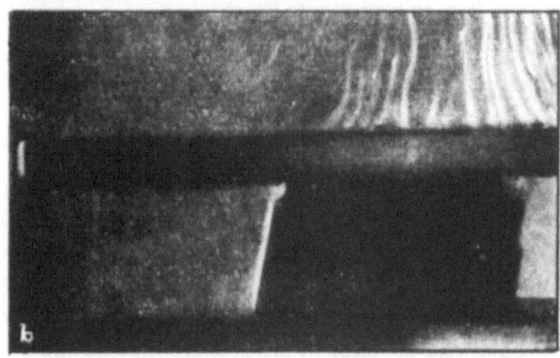

Fig. 10. Alum crystal growing between two glass plates (a) at the start, and (b) at the end. The white band to the left of the face is the growth layer. The band bends at the top in (b) because this part of the face has grown hardly at all. × 3.

was lifted less than the one on a cube face (Fig. 7). The converse also often occurred, even when the conditions were exactly the same. It is clear that the effects with spheres 1-5 mm in diameter are determined by purely local (partly uncontrolled) growth conditions.

It is clear now that the pressure measured in this way is not the crystallization pressure (as defined above), but is some effective pressure that acts when a crystal traps or repels a foreign particle during actual growth. Hence, there is no contradiction between Correns' and Shubnikov's results, and no explanation in terms of differing supersaturations is needed, especially since Correns' data show that the pressures that should have been found at Shubnikov's supersaturation were about 500 g/cm^2, and not about 0.14 g/cm^2.

Bokii. We were not able to measure crystallization pressures by Bokii's method. We found, in agreement with [5], that, if the plate (here a glass one) that covered the top face was larger than the face, we could produce no growth funnel. The top face did not grow under these conditions, whereas the lower one did, and formed a growth funnel. The cause is clearly the difference in the supply of material. We checked this by Toepler's method [13] (on which a shadow pattern is produced), and used conditions similar to those used by Bokii. Figure 10 shows two photographs, of which (a) was taken at the start of growth, while (b) was taken after a certain interval. It is clear that even the open face failed to grow near the top plate. Normal stirring failed to alter the picture very much. Thus the supply conditions determine whether the crystal traps or repels an obstacle. This topic will be dealt with in detail elsewhere.

Bokii's results can be explained if we assume that the top face grows at a very low supersaturation (because the supply of material is poor), in fact, at one much below the mean value in the solution. Then Bokii's results agree with Correns', and not with Shubinkov's, as Bokii points out [11].

SUMMARY

1. The crystallization pressures Correns and others have measured for the (111) face of alum (about 20 kg per cm^2 at a supersaturation of 1.2) are confirmed as to order of magnitude.

2. The supply of material determines whether a crystal traps or repels an obstacle.

I wish to thank Academician Shubnikov and Dr. Sheftal' for guidance in this work, and E. M. Akulenok for assistance with the experiments.

18

LITERATURE CITED

[1] C. Correns and W. Steinborn, Z. Krist. A101, 117-133 (1939).

[2] G. Becker and A. Day, Proc. Washington Acad. Sci. 7, 283-288 (1905).

[3] K. Andree, Geol. Rund. 111, 7-15 (1912).

[4] W. Bruhns and W. Mecklenburg, 6th Jahresbericht d. Niedersächsischen Geolog. Vereins 92-115 (1913).

[5] S. Taber, Amer. J. Sci. 532-556 (1916).

[6] C. Correns, Sitz. Ber. Preuss. Akad. Wiss., Phys.-math. 11, 1 (1926).

[7] A. Shubnikov, Z. Krist. 88, 466-469 (1934).

[8] A. V. Shubnikov, Trudy Lomonosov Inst. Akad. Nauk SSSR 17-21 (1935).

[9] E. Scheil, Z. Metallkunde 27, 4, 76 (1935).

[10] I. N. Fridlyander and N. A. Vysotskaya, Doklady Akad. Nauk SSSR 62, 1, 71-3 (1948).

[11] G. B. Bokii, Trudy Inst. Kristall. Akad. Nauk SSSR No. 5, 143-8 (1949).

[12] P. S. Vadilo, Zap. Kishinevsk. Univ. 3, 1, 181 (1951).

[13] N. A. Valyus, Solution Optics [in Russian] (Gostekhizdat, 1949).

THE GROWTH CONDITIONS OF CRYSTALS IN CONTACT
WITH LARGE OBSTACLES

V. Ya. Khaimov-Mal'kov

The repulsion or capture of foreign particles encountered by a growing crystal has two aspects: firstly, the pressure the crystal may exert on the body, and secondly, the supply of material to areas in contact with such body.

We have dealt already with crystallization pressure. It has been shown theoretically [1]* and experimentally [1, 2] that at small supercoolings (~ 0.5°C) and supersaturations (~ 1.1) the pressure may be fairly large (~ 8 kg/cm²). This implies that almost any particle should be pushed away. We have so far not dealt with the supply of material to areas covered by foreign bodies. The supply conditions in fact largely determine whether capture or repulsion occurs.

It is found that, although the crystallization pressure is large, the body is often not pushed away. Figure 1a shows, for example, that, at a supercooling of ~ 0.1°C, abenzophenone crystal will repel lycopodium particles. If the supercooling rises to 0.5-1°C, the crystal starts to form inclusions (the tracks behind the particles are seen in Fig. 1b) [7]. It is clear that some particles have already entered the crystal. Almost all the particles are trapped at larger supercoolings. The particles are again repelled if the supercooling is reduced to 0.1°C again (Fig. 1c).

It would thus appear that the probability of trapping increases with the supercooling (supersaturation), although the crystallization pressure increases with the supersaturation (supercooling). Some have in consequence concluded that the pressure falls as the supersaturation rises [4, 5]. These facts do not run counter to the theory, however. They can be explained in terms of the reduced growth rate sometimes found on areas covered by obstacles. We now have to consider how the rate of crystallization is affected by the crystallization parameters (supersaturation, rate of stirring, etc.).

Theory shows that the pressure cannot depend much on the natures of the particle, of the melt (solution), and of the crystal [3]. We know, however, that a (111) face in alum repels a glass plate, whereas a (100) face sticks to such a plate. Both faces stick to a mica plate [1, 6]. The facts give the impression that the pressure may be caused by surface-tension forces.

These facts also can be explained in terms of the differing supply conditions at covered faces. A covered face is fed via a thin film of solution (melt) between the body and crystal; in some cases this film's properties depend on the face, the body and the solution, or else the supply mechanism depends on them. We must consider how the growth rate of a covered face depends on the face, on the nature of the body, on any surface-active impurities that may be present, and so on.

We give here some experimental evidence on the supply conditions at covered faces.

Experimental Results on the Growth of Covered Faces

Methods. We have used the following effect to determine how the supply conditions depend on the crystallization parameters. A small crystal (seed) in a supersaturated solution has some one face in contact with a smooth surface (of glass, mercury, mica, or of the mother-liquor itself); soon the seed starts to rise above the surface, because the face at the bottom grows. A growth funnel [6, 9] forms on this bottom face (Figs. 2 and 3). Now the bottom face is subject to a force that increases as the crystal grows. By definition the crystallization pressure is the pressure at which growth ceases, which pressure is reached when the weight of the crystal is large enough. Here

* See my paper in this volume, p. 3.

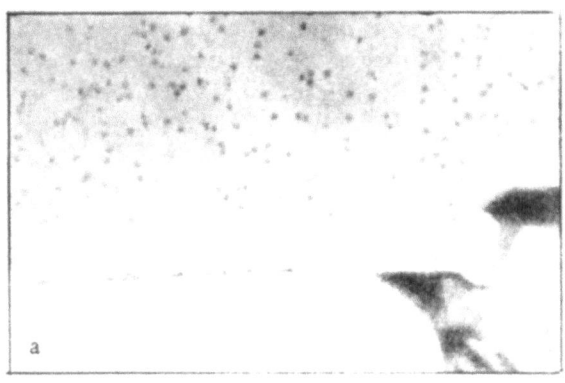

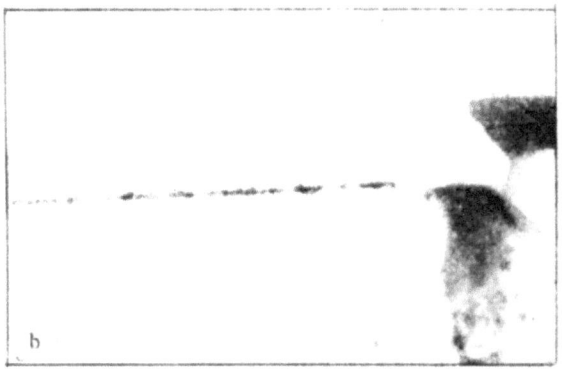

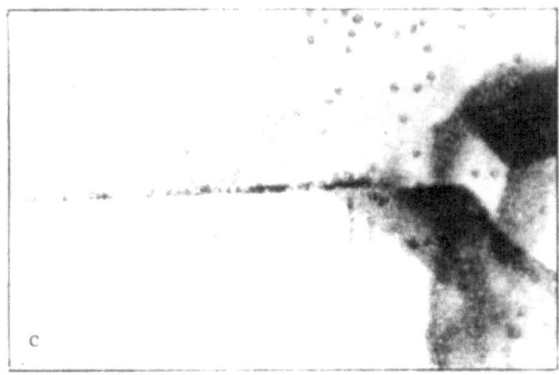

Fig. 1. Trapping and repulsion of lycopodium particles by a growing benzophenone crystal at various supercoolings: a) ∼ 0.1°C; b) 0.5-1°C; c) supercooling reduced to 0.1°C again. The top half shows the melt, with the particles floating in it; the bottom half shows the growing crystal (a and b × 40; c × 75).

the crystal acts as a foreign body. The lower face is fed via a thin film of solution between the crystal and the support. We have determined the growth rate of the covered part of the face from a graph we have constructed to show the relation of height raised h_1 to time t_1. We have taken that the open faces grow at a fixed rate, and the calculation is easy if we know the length l_1 corresponding to any h_1, the total time t_0 of the experiment, and the corresponding distance l_0 (Fig. 3). If the initial and final weights of the crystal, P and P_0, are also known, we can find the weight P_1 corresponding to any height h_1. Now h_1 and l_1 were measured at the end of the experiment with an MIS-11 microscope (without changing the setting of the crystal); the accuracies were 10 μ for h_1 and 25 μ for l_1. The results gave, nonetheless, only a semiquantitative relation between P_1 and h_1. The cause is the fluctuations in the growth conditions at the bottom face. Figure 4 shows how reproducible were the

Fig. 2. Growth funnel on the bottom (111) face of an alum crystal.

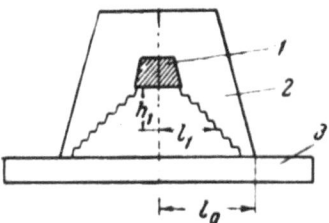

Fig. 3. Growing crystal with a growth funnel on its lower face: 1) seed; 2) crystal, 3) support.

results for two crystals lying on octahedron faces. The growth rate v_1 of the covered face was found by differentiating the $h_1 = f(t_1)$ curve graphically, which itself results in a large scatter. Wherever possible we have given the direct (undifferentiated) data.

We used potash alum. The temperature was kept constant to ± 0.025°C. The supersaturation of the large volume (4 liters) of solution did not vary by more than 0.26%; the concentration was measured in terms of the refractive index. The surfaces to be used (glass, mica, etc.) were cleaned carefully, and were set level in the crystallizer. It is correct to determine the time t_1 for a covered face from $t_1 = l_1/v_0$ (where v_0 is the growth

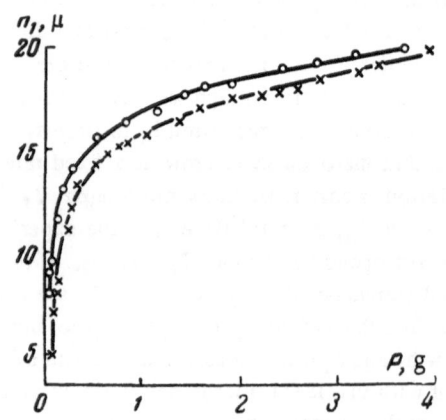

Fig. 4. Relation of height raised on a (111) face to weight.

rate of an open face; $v_0 = l_0/t_0$), or the weight P_1 from $P_1 = P_0 l_1^3/l_0^3$ only if the crystal grows uniformly in all directions. We checked this by examining the surfaces of the negative pyramid on the bottom face (Fig. 2) for flatness. This is a very sensitive test for changes in the growth conditions. Figure 5 shows a completely transparent crystal, but steps are clearly visible on the pyramid, and the number of steps corresponds to the number of changes in temperature exactly. The temperature was reduced in steps once every 1-2 days (see table).

Relation of growth rate for a covered face to supersaturation and weight of obstacle. Figure 6 shows typical curves relating v_1 to supersaturation C/C_0 for a fixed weight P_1; Fig. 7 shows v_1 vs. P_1 with $C/C_0 = $ const. We used crystals lying on their (111) faces on optically flat glass. The solution was stirred at about 20 rpm.

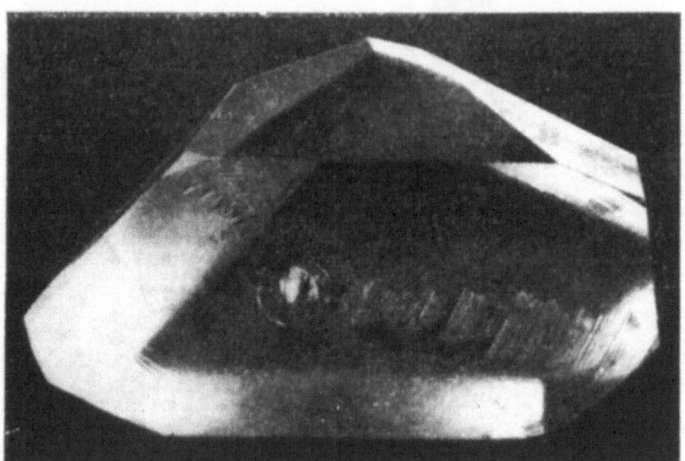

Fig. 5. Lower face of a crystal with its pyramid. The number of steps corresponds to the number of changes in temperature.

Temperature-Time Data

Time		t, °C	Time		t, °C
day	hour		day	hour	
9/X 1957	17	42.1	16/X 1957	17	41.3
10/X 1957	17	42.0	17/X 1957	17	41.0
12/X 1957	17	41.9	18/X 1957	17	40.7
14/X 1957	17	41.7	20/X 1957	10	40.0
15/X 1957	17	41.5			

The curves show that:

1) the growth rate for a covered face increases with the supersaturation;

2) the growth rate for a covered face decreases as the weight increases at any fixed supersaturation;

3) this latter fall in rate is especially marked at high supersaturations, but is slight at low ones; i. e., that the growth rate increases very rapidly with supersaturation at small loads, and fairly slowly at high loads.

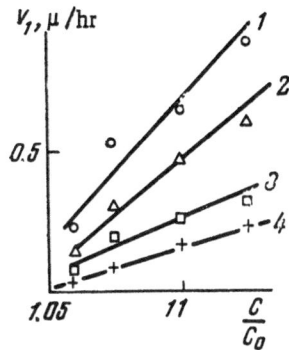

Fig. 6. Relation of growth rate for a covered face to supersaturation: 1) $P_1 = 0.043$ g; 2) $P_1 = 0.164$ g; 3) $P_1 = 0.152$ g; 4) $P_1 = 1.200$ g.

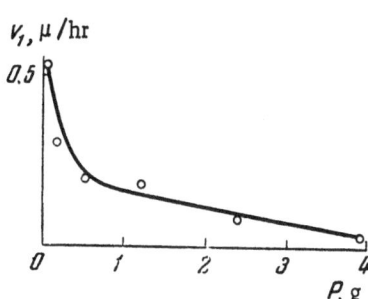

Fig. 7. Relation of growth rate for a covered face to load at $C/C_0 = 1.075$.

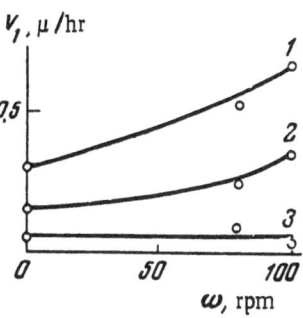

Fig. 8. Relation of growth rate for a covered face to rate of stirring for crystals of various weights at $C/C_0 = 1.075$: 1) $P_1 = 0.05$ g; 2) $P_1 = 0.40$ g; 3) $P_1 = 1.45$ g.

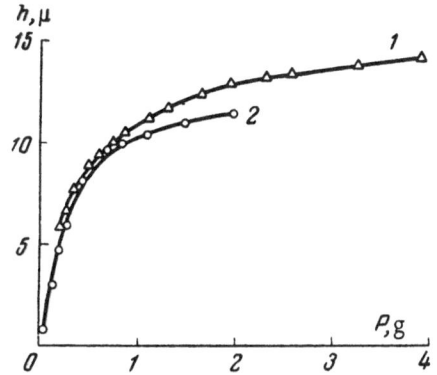

Fig. 9. Relation of growth rate for a covered face to a supersaturation at $v_1 = 0.125$ mm/hr: 1) $C/C_0 = 1.11$; 2) $C/C_0 = 1.07$.

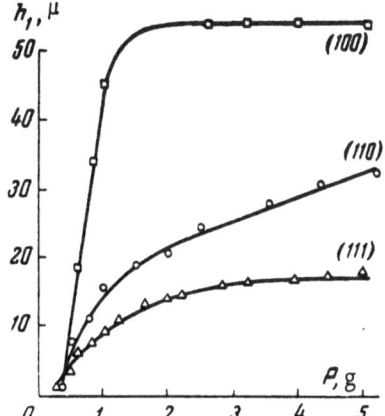

Fig. 10. Relation of height raised to weight of crystal for a supersaturation of 1.06. Crystal growing on optical glass.

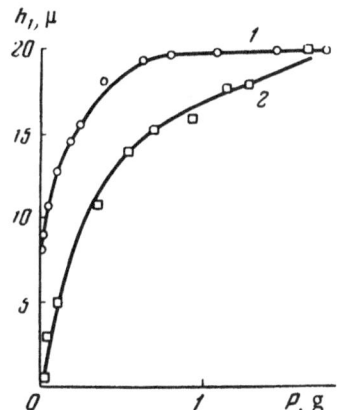

Fig. 11. Relation of height raised to weight of crystal for a (111) face and various supports: 1) mica, 2) optical glass.

Rate of stirring for crystals of various weights. Figure 8 shows curves for v_1 vs. ω (where ω is the angular velocity of the stirrer) for loads of fixed weights, and v_1 vs. P_1 at $\omega = $ const. The tests were done under the same conditions as were those for v_1 vs. C/C_0 (except for the stirrer); it is clear that the stirrer, which turned fairly rapidly, affected the reproducibility, especially with the small crystals, which might have been disturbed.

Figure 8 shows that the growth rate increases with the rate of stirring for small crystals, but that any such increase is very small for large crystals.

Figures 6 and 8 shows that the relations of growth rate to supersaturation and rate of stirring are such that, if the loads are equal and the growth rates of the open faces are equal, the covered faces grow more rapidly at the higher supersaturations. The open faces are made to grow at equal rates by increasing either the supersaturation or the rate of stirring, as may be appropriate. Now v_0 is the same in both cases (Fig. 9), i. e., the time scales are the same, so the ratio of growth rates for covered faces can be deduced for any given load.

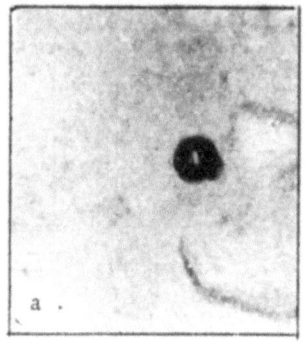

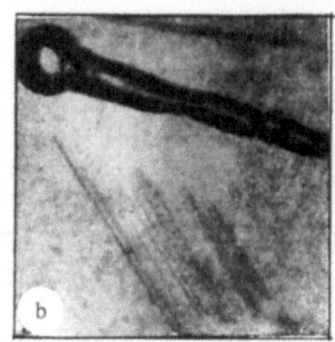

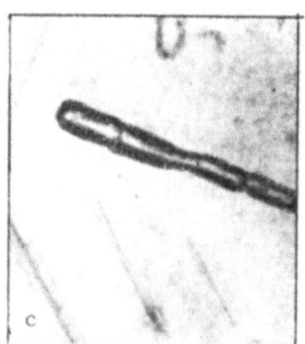

Fig. 12. Accelerated growth of a
face in a thymol melt when an attached
air bubble is present: a, b) × 12;
c) × 9.5.

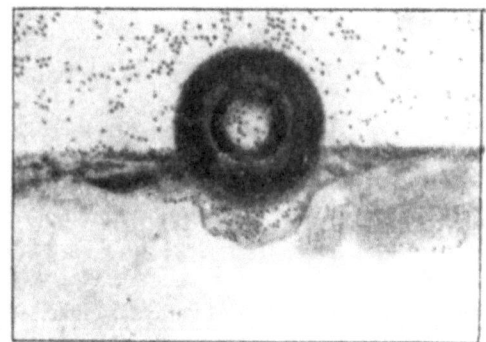

Fig. 13. Face of a thymol crystal repelling an air
bubble when the two are separated by a layer of
lycopodium particles. A negative crystal is seen
under the bubble. × 20.

Relation of supply conditions for a covered face
to material of support and to structure of the face. Figures 10 and 11 show how h_1 varies with P_1 for different faces and supports (glass, mica). The tests were done simultaneously in a single crystallizer, and the final weights were similar, and so we can assume that P_1 is proportional to t_1^3 for all curves, i. e., we can compare the growth rates of different covered faces. It is clear that the rates depend very much on the faces and on the material (obstacle). The effect is especially marked at high loads, but is slight for crystals weighting ~ 0.1 g. The $h_1 = f(P_1)$ curves for (100) against glass and (111) against mica are especially interesting; it has been claimed [1, 6] that these faces stick to these materials. Figures 10 and 11 show that the faces do not stick at once, but only when some definite weight has been reached. The center of any such face always shows a small funnel, followed then by a flat face.

It is very common to find a body of any weight and size sticking to a face. Figure 12 shows an air bubble sticking to the face of a thymol crystal. A simple test showed that the supply conditions, structure of the face, etc., are here responsible for the effect, whereas the crystallization pressure is not. Thymol repels lycopodium readily, as does benzophenone (Fig. 1). Lycopodium particles between the bubble and a face to which the bubble has previously stuck block the face and cause it to repel the bubble (Fig. 13). A negative crystal is seen under the bubble.

Figure 14 shows another interesting case. When a benzophenone crystal in a melt meets drops of mercury, we get new faces formed, which faces alter the supply conditions under the drops (the drops start to be distorted). The crystal grows on once the drops have been surrounded by these new faces.

Fig. 14. New faces forming when a benzo-phenone crystal in a melt meets drops of mercury. × 30.

Conditions for an Obstacle to be Trapped Mechanically

The above examples show that body, solution (melt) and crystal interact in three main ways. In the first, the covered face is supplied continuously, i. e., there is always a layer of solution or melt between the body and the crystal; in the second the supply is partly cut off, and the layer exists only up to some limiting weight of body; and in the third, the body sticks at once to the face, no matter what its weight. It is clear that the conditions dealt with above are not the only ones that determine the result. The body may be trapped purely mechanically, simply because the growth rate of the covered and open areas on the face (v_1 and v_2, respectively) are very different. The ratio v_1/v_0 serves to indicate to what extent the trapping is purely mechanical.

Figure 15 shows v_1/v_2 vs. supersaturation for various loads (optical glass and a (111) face). The curves give rise to the following conclusions.

1. The ratio v_1/v_0 falls as the supersaturation rises, i. e., trapping becomes more probable.

2. The ratio v_1/v_0 is larger for light particles and smaller for heavy ones, at any given supersaturation, i. e., trapping becomes more probable as the weight increases, if C/C_0 is constant.

3. The difference between the v_1/v_0 values for light and heavy particles decreases as the supersaturation rises.

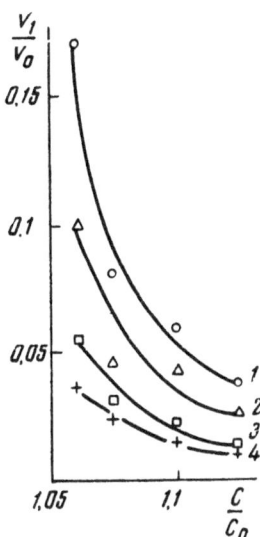

Fig. 15. v_1/v_0 vs. supersaturation for various loads (optical glass and a (111) face): 1) $P_1 = 0.043$ g; $v_0 = 0.0137$ mm/hr ($C/C_0 = 1.060$); 2) $P_1 = 0.164$ g; $v_0 = 0.0652$ mm/hr ($C/C_0 = 1.075$); 3) $P_1 = 0.520$ g; $v_0 = 0.1100$ mm/hr ($C/C_0 = 1.100$); 4) $P_1 = 1.200$ g; $v_0 = 0.2300$ mm/hr ($C/C_0 = 1.125$).

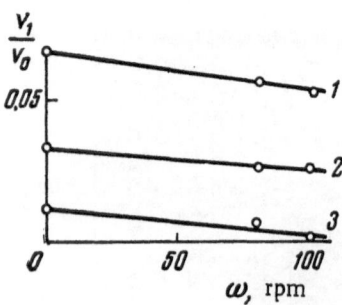

Fig. 16. v_1/v_0 against rate of stirring for various loads: 1) $P_1 = 0.05$ g, $v_0 = 0.045$ mm/hr ($\omega = 0$ rmp); 2) $P_1 = 0.4$ g, $v_0 = 0.097$ mm/hr ($\omega = 80$ rpm); 3) $P_1 = 1.45$ g, $v_0 = 0.125$ mm/hr ($\omega = 100$ rpm).

Fig. 17. Transfer of an air bubble from one face of a thymol crystal to another. Crystal growing from the melt. × 15.

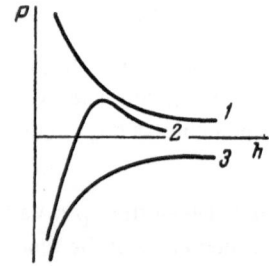

Fig. 18. Relation of thickness of a layer between a face and an obstacle to pressure applied.

Figure 16 shows v_1/v_0 against rate of stirring for various loads (optical glass and a (111) face). The curves show that v_1/v_0 falls somewhat for heavy particles, and scarcely alters for light ones, as the rate of stirring increases.

Figures 15 and 16 imply that v_1/v_0 increases with the supersaturation when v_0 is constant, i. e., a foreign body is more likely to be trapped, at any given growth rate, at points where the stirring is most vigorous and the supersaturation least. Thhs conclusion applies if the particles are at rest and the solution (melt) is stirred.

Thus, if we are to assess the effects of any factor on mechanical trapping, we must know how this factor affects the growth rate of the open areas as well as how it affects the rate in the covered areas. This is true not only for parameters such as the supercooling (supersaturation), stirring rate, impurity concentration, face, etc., but also for the obstacle itself. In some cases the obstacle affects both v_1 and v_0. Figure 12 shows a growing thymol crystal. A bubble stuck to one face may increase the rate of growth of that face relative to the rates for the other faces. The face regains its old rate if the bubble is lost. This change in rate may (if other conditions stay the same) cause a face that did not trap particles to begin to do so when it acquires a bubble. The bubble may quite often be transferred from one face to another (adjacent) one, because fast-growing faces soon become small. The bubble is transferred from the residual point-face to a nearby one, which causes a sharp turn in the inclusion (Fig. 17a, b). The angle of the turn must nearly or exactly equal the angle between the face-normals.*

DISCUSSION

Our results show that we supposed correctly that the supply conditions at the covered areas are determined mainly by the properties of the thin layer supplying such areas. We find in fact that the resemblance between our layers and some other thin layers is very close, even down to quantitative details. Figure 18 shows how the thickness of a surface layer depends on the pressure applied [10-12]. The curves have the same general shape as those relating h_1 or v_1 to P_1 (Figs. 2 and 7); the region where the thickness varies rapidly with pressure (about 1 g/cm² for water [10]) is the same as the one where h_1 or v_1 varies rapidly with P_1. The effects of the surfaces, of the liquid, of impurities, etc., on the thickness [11] appear only at fairly high pressures, as in the case of our supply layers, where $h_1 = f(P_1)$ is affected by the face, by the support, by the impurities, etc., only when the weight of the crystal is fairly large.

A thermodynamic criterion of "thinness" has been given in terms of the sign of $\partial\sigma/\partial h$ [12], where σ is surface tension and $\underline{h}$ is thickness; the result agrees with our

---

* The way in which such inclusions are formed may be judged from the shapes.

data on the thickness of the thin layer. A thin layer is stable at any pressure if $\partial\sigma/\partial h < 0$ (Fig. 18, 1). This corresponds to a stable supply layer, e. g., on a (111) face in alum in contact with a glass plate (Fig. 10). The thin layer is unstable at any pressure if $\partial\sigma/\partial h > 0$, and the surfaces are drawn together (Fig. 18, 3); this corresponds to the case where the obstacle always sticks to some face, e. g., air bubbles to thymol crystals (Fig. 12). But if $\partial\sigma/\partial h$ changes in sign with pressure, the layer is stable up to some pressure, past which sticking occurs (Fig. 18, 2). This case may correspond to a (100) face in alum in contact with a glass plate, where the face grows only up to some limiting weight of crystal, and thereafter sticks tightly to the glass (Fig. 10).

We can explain how the growth rate of a covered face, and v_1/v_0, depend on supersaturation (supercooling), stirring, and weight of crystal in terms of how diffusion or flow in thin layers depend on these same parameters [3, 13]. Concentration or convection flows are much more important than diffusion in transferring material in a liquid [14]. This mode of transfer determines the growth rates of open faces. Growth on covered faces is determined mainly by diffusion in thin supply layers. Hence, improved stirring increases the growth rate on open faces, but has little effect on closed faces (Fig. 8, $P_1 = 1.45$ g), so the ratio v_1/v_0 falls (Fig. 16, $P_1 = 1.45$ g). Now the flow in the layer ($\vec{J_1} \sim D_1$ grad c, where D_1 is the diffusion coefficient and $\underline{c}$ is concentration) depends on the supersaturation, so the growth rate on a covered face is more dependent on supersaturation than on stirring. The growth rate must be roughly proportional to the concentration gradient, as is in fact found (Fig. 8).

Hence, the supply conditions on covered faces are determined by the physical and chemical properties of thin layers.[*]

The above argument now enables us to separate out three aspects of the problem of how a growing crystal repels foreign particles.

1. The crystallization pressure the crystal can exert. This is fairly large at low supersaturation (supercoolings), is almost independent of the nature of the particle and solute, and depends on the energy of the phase change. Any foreign particle should be rejected by a growing crystal.[*]

2. The areas of a face in contact with particles are supplied via physically thin layers. The properties of such layers, and hence the growth conditions, are determined by the way the surface tension is related to parameters such as thickness (which is a function of the pressure applied), temperature, concentrations of solute and impurities, structure of face, nature of solute, etc. The relationship in a final reckoning determines how the growth rate of a covered area is related to surface tension, up to the limit where the particle sticks to the face and the growth rate is zero.

3. Even if the crystallization pressure is substantial, and the covered face is supplied with material, the crystal may trap the particle. Here the result is determined by the ratio of the growth rates on the covered and open faces. This ratio determines how long it takes for the crystal to trap the particle, which time varies with v_1 and v_0.

SUMMARY

1. Data are given for alum on the relation of growth rate on a face covered by a macroscopic obstacle to supersaturation, rate of stirring, nature of face and material of obstacle.

2. The data show that the properties of the thin layer of solution between the face and the body determine whether that face will capture that body under any given conditions.

3. A picture of how a crystal engulfs foreign particles has been presented and, in part, discussed.

I wish to thank Academician Shubnikov and Dr. Sheftal' for guidance in this work.

LITERATURE CITED

[1] C. Correns and W. Steinborn, Z. Krist. A 101, 117 (1939).

[2] E. Scheil, Z. Metallkunde 27, 4, 76 (1935).

---

[*] It is very difficult to study thin layers directly. We can use the effect described above (in which separate very slight effects are summed) in order to study some properties of thin layers.

[**] Crystallization in and on porous solids (concrete, brick, etc.) are special cases of this.

[3] S. V. Nerpin and B. V. Deryagin, Doklady Akad. Nauk SSSR 100, 1 (1955).

[4] N. N. Fridlyander, Trudy VIAM, 1946.

[5] P. S. Vadilo, Uch. Zap. Kishinev. Univ. 3, 1, 181 (1951).

[6] G. B. Bokii, Trudy Inst. Kristall. Akad. Nauk SSSR 5, 143 (1949).

[7] A. V. Shubnikov, How Crystals Grow [in Russian] (Izd. AN SSSR, 1935) p. 131.

[8] C. Correns, Sitz. Ber. Preuss. Akad. Wiss., Phys.-math. Kl. 11, 1 (1926).

[9] S. Taber, Amer. J. Sci. 532 (1916).

[10] B. Deryagin and E. Obukhov, Colloid J. (USSR) 1, 5, 385 (1935).

[11] B. Deryagin and M. Kusakov, Izv. Akad. Nauk SSSR, Ser. Khim. No. 5, 741 (1936).

[12] Ya. I. Frenkel', The Kinetic Theory of Liquids [in Russian] (Izd. AN SSSR, 1945) p. 305.

[13] B. Deryagin, Colloid J. (USSR) 17, 3, 207 (1955).

[14] V. G. Levich, Physicochemical Hydrodynamics [in Russian] (Izd. AN SSSR, 1952).

[15] S. V. Nerpin and B. V. Deryagin, Doklady Akad. Nauk SSSR 100, 1 (1955).

THE CRYSTALLIZATION OF ALUM FROM WATER IN
AN ULTRASONIC FIELD

A. P. Kapustin and V. E. Kovalyunaite

In the last few years several papers have appeared on the effects of ultrasonics on crystallizing substances [1, 2]. These papers dealt with polycrystalline materials formed from melts, and did not touch on monocrystals. We give here some data on the effects of ultrasonics on monocrystals of potash alum.

Growth of Monocrystals in an Ultrasonic Field of Frequency 30 kc

A preliminary test of the effects of ultrasonics on the growth of alum monocrystals in a slightly supersaturated solution [3] showed that the linear growth rate of the octahedron face was increased at a frequency of about 2 Mc. The wavelength in the solution was short, and so it was difficult to distinguish the effects of nodes and antinodes; in this work we used a frequency of about 30 kc, which corresponds to a wavelength of about 5 cm.

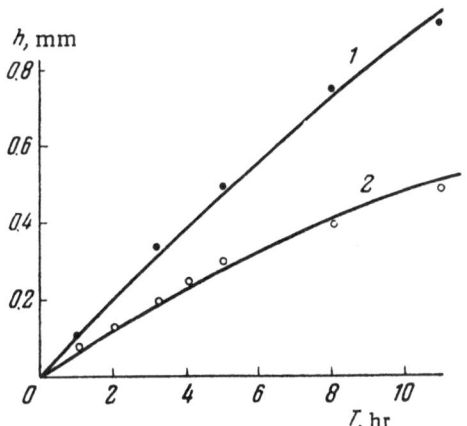

Fig. 1. Growth rates of octahedron faces:
1) at an ultrasonic antinode; 2) no ultrasound.

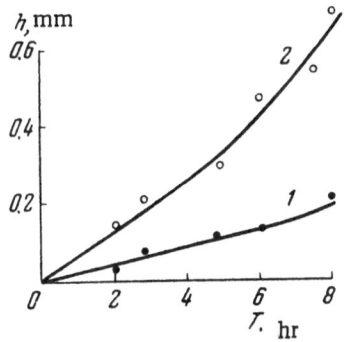

Fig. 2. Growth rates of octahedron faces:
1) at an ultrasonic node; 2) no ultrasound.

We used alum crystals 6-8 mm in size, mounted on wires. The solution was held in two rectangular tanks with double plexiglas walls, tightly closed with lids. The crystals were set at the same height in each, with the octahedron faces parallel to the bottom in every case. One tank was put directly on the magnetostriction source. The lower octahedron face was then parallel to the wavefronts, while the side faces were inclined. (The positions of the nodes and antinodes had previously been located.) The crystal could be moved to different parts of the tank. The tanks were filled with a warm solution of alum, of concentration such that the supersaturation was small when the solution had cooled. The temperature was controlled by a thermostat. The intensity at no time exceeded 0.1 w/cm^2.

At first, we used crystals whose sizes were much smaller than the wavelength. Figure 1 shows the absolute increase in size for the octahedron face of two crystals, one of which was located at an antinode, and the other of which grew under normal conditions. The mean slope of the curve (the growth rate) was almost doubled by the ultrasonics. The masses of the crystals were almost equal (a difference of 10-15 mg in a total increase of 200-300 mg), because the octahedron faces that were not parallel to the wavefronts were less well developed.

Figure 2 relates to lower octahedron faces placed at node positions. Here the ultrasonics reduced the growth rate by a factor of 2-3. The masses of the crystals were almost equal, because the inclined faces were more highly developed.

We then used monocrystals whose sizes were comparable with the wavelength. A crystal was cut parallel to the octahedron faces into a plate, and was then set with its plane at 90° to the wavefronts. The top and bottom edges were at nodes, and the middle of the face was at an antinode. At high intensities the crystal dissolved 7-10 times more rapidly at the antinode than it did at the nodes. There was no appreciable effect on the growth, except that the surface became coated with small crystals at the antinode. There were less such crystals at the nodes. The faces of a crystal grown away from the ultrasonics were completely smooth.

Production of Crystallization Nuclei in the Solution at 30 kc

The ultrasonics produced a host of crystallization centers almost at once (Fig. 3a, b) when a highly supersaturated solution (one at 6-8°C below its saturation temperature) was exposed to the waves. This effect was found at 700 kc also.

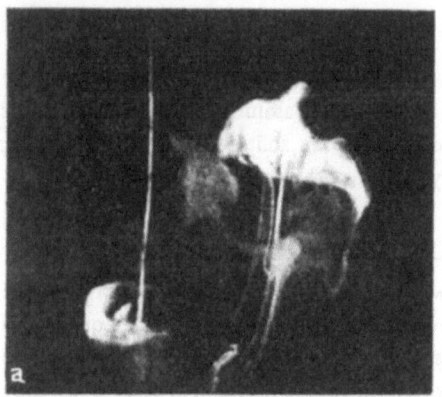 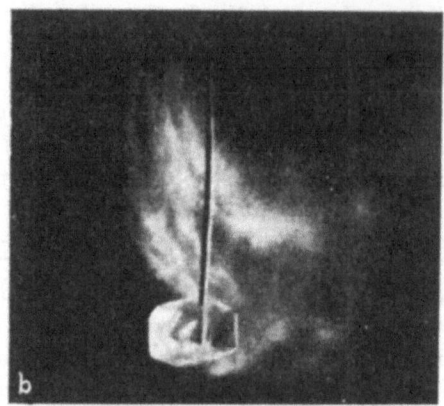

Fig. 3. Some phases in the formation of a cloud by the ultrasonic waves.

The crystals were at first so numerous that the solution became opaque. After 4-5 min it was possible to see small crystals settling out slowly. After 10-12 min the solution was again clear. The saturation temperature of this clear solution was, at most, 1-2°C above room temperature.

The effect is very dependent on the intensity, on the concentration, and on the position of the seed in the tank. Below 0.2 w/cm² the nuclei formed as a cloud only at large supercoolings (6-8°C); at high intensities a supercooling of about 1°C was enough. The cloud appeared and developed the more rapidly the higher the intensity and the supersaturation. The nuclei were produced more rapidly if the crystal was moved from an antinode to a node. It was important to have stable standing waves in the solution in order to produce the effect.

At a certain supersaturation a bunch of small crystals appeared at the antinode next to the antinode occupied by the crystal; larger crystals could be seen to develop, and smaller ones to appear in nearby areas. The bunch of crystals fell to the bottom when the intensity was increased. A large crystal inserted in such a solution acquired some of the small crystals if it was set normal to the waves. We found that most of the little crystals stuck to the large one were in fact forming a polysynthetic parallel growth with it.

This result confirms Shaskol'skaya and Shubinkov's findings [4]. Our small crystals were, however, very much smaller, because we gave them no time to grow. The result also confirms Keck's opinion [5] (which has been confirmed by experiment [6]) that heavy particles accumulate at the antinodes. This gives us the reason why the octahedron face parallel to the wavefront at an antinode grew so rapidly.

Concentration Currents

One may see how the ultrasonic waves affect the supply of material to the crystal by viewing the concentration currents. A crystal was placed at a pressure node (at 30 kc); the growth currents were straight and were almost unaffected when the ultrasonic waves were switched on (Fig. 4a). The currents were disturbed at an antinode (Fig. 4b). The nodes can in fact be detected from the behavior of the concentration currents. It may be that the disturbance to the currents improves the supply of material to the crystal.

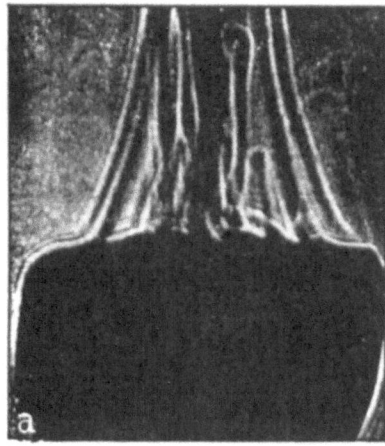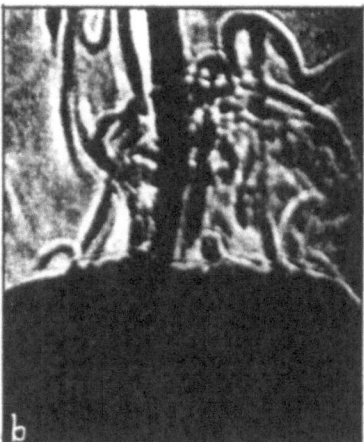

Fig. 4. Concentration currents near a growing crystal: a) at a pressure node;
b) at an antinode.

SUMMARY

1. The growth rate on an octahedron face of an alum crystal is dependent on whether the face is at a node or at an antinode. The total increase in weight is not much affected, though. At high intensities a crystal dissolved more rapidly when it is at an antinode.

2. Ultrasonics can cause crystals to form en masse in a highly supersaturated solution of alum. The time needed to cause this effect depends on the intensity and on the supersaturation; frequency is not found to have any effect.

3. Some of the small crystals produced by the ultrasonics stick to the large crystal, and most form parallel polysynthetic growths.

4. It has been confirmed that heavy particles migrate to pressure antinodes.

5. The concentration currents at nodes differ from those at antinodes.

LITERATURE CITED

[1] A. P. Kapustin, Trudy Inst. Krist. Akad. Nauk SSSR No. 10 (1954).

[2] E. Hiedemann, J. Acoust. Soc. Amer. 26, 5 (1954).

[3] A. P. Kapustin and V. E. Kavalyunaite, Kristallografiya 1, 6 (1956).

[4] M. Shaskol'skaya and A. V. Shubnikov, Z. Krist. 85, 1 (1933).

[5] G. Keck, Acustica 5, 2 (1955).

[6] Kh. S. Bagdaserov, Kristallografiya 3, 4, 487 (1958).

Institute of Crystallography, Academy of Sciences, USSR

THE ADSORPTION OF THIONIN BLUE BY GROWING LEAD NITRATE
CRYSTALS, AND THE EFFECTS ON THE MORPHOLOGY OF
THE CRYSTALS

E. N. Slavnova

Organic compounds are usually surface-active, and affect crystallization processes by adsorption. A growing crystal has surfaces that are physically different and that are not constant in size, and the crystallization conditions also have certain effects; so the behavior and ultimate fate of any adsorbed material depend on many factors.

Vedeneeva and I have used optical methods (especially spectrophotometry) to study adsorption effects [1].

We used thiazine dyes and the nitrates of lead, barium and strontium as our standard materials. Our first studies were on the state of aggregation of the impurity in the host crystal. A later study was on the effects of adsorbed dyes on the morphology of the crystals.

There have been many papers on the effects of impurities on crystal morphology, both early ones, which were phenomenological in type [2-5], and later ones, which deal with the mechanisms whereby impurities affect habit [6-9]. Whetstone's work [10-13] is of special interest, because he proposed a mechanism for the effects of organic dyes on habit for ionic crystals. The predictions in some cases agree well with experiment; the mechanism derives from the correspondence between the structure of the crystal and the polar groups in the dye.

We have sought to determine how the morphology is related to the concentration of the dye in the mother liquor. I have dealt with this topic for methylene blue already [14-17]. The solvent was allowed to evaporate slowly under steady conditions from a solution of lead nitrate containing methylene blue; it was found that the color and habit were highly dependent on the initial concentration of dye in the solution. It was also found that traces $(3.0 \cdot 10^{-3} \%)$ of dye were absorbed maximally. Higher $(1.0 \cdot 10^{-2} \%)$ concentrations showed an upper bound to the uptake; above this bound the crystals grew uncolored. Barium nitrate also shows a maximum uptake of methylene blue.

These effects had not been found by earlier workers for lead (or barium) nitrate and methylene blue [18-24], and so it may be that the maximum uptake is caused by certain conditions during the crystallization. In the present work (which is being continued) we have used thionin blue, a dye related to methylene blue, which is also taken up well by lead nitrate crystals, and have given this aspect special attention. The cause of these maxima is of particular interest, because Buckley [25] has pointed out that many dyes show adsorption maxima, the reasons for which have not yet been found.

Here we have examined the effects of initial thionin blue concentration on the form, transparency and coloring for twenty concentrations in the range $1 \cdot 10^{-3}$ to $5 \cdot 10^{-1} \%$. We looked very carefully for possible maxima in uptake, and examined the coloring of the crystals.

The crystals were either microscopic ones observed during growth, or the ones produced from 25 g of solution. The solutions were made by adding lead nitrate to aqueous solutions of the dye, previously made up at appropriate concentrations.

Results for Monocrystals

A small drop of solution was allowed to evaporate slowly on a glass slide; the crystals that grew were viewed in polarized light under the microscope until the drop dried up completely. The dye concentration varied during this evaporation, but the effect of the initial concentration on the morphology was quite clear and reproducible.

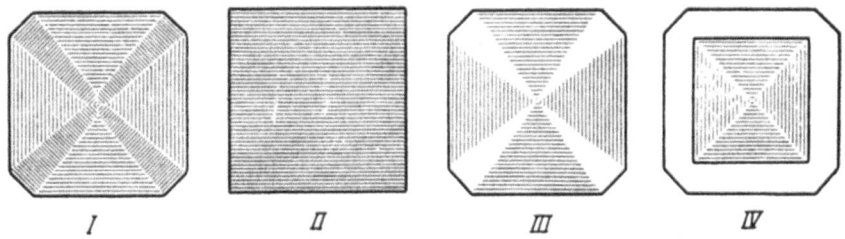

Fig. 1. The effects of thionin blue concentration on the morphological features of lead nitrate crystals. Dye concentrations: I) $(2-5) \cdot 10^{-3}\%$; II) $(1-4) \cdot 10^{-2}\%$; III) $(5-10) \cdot 10^{-2}\%$; IV) $(2-5) \cdot 10^{-1}\%$.

Fig. 2. Lead nitrate crystals containing thionin blue. Dye concentrations: a) $3 \cdot 10^{-3}\%$; b) $2 \cdot 10^{-2}\%$; c) $1 \cdot 10^{-1}\%$; d) $2 \cdot 10^{-2}\%$.

We were able to assign the crystals to four groups in accordance with the morphologic features. The results are given in the table and in Figs. 1 and 2.

The first group is made up of those crystals formed at dye contents of $(2-5) \cdot 10^{-3}\%$. These crystals have the dye selectively absorbed by the cube faces. The growth pyramids of the tetrahedron faces stayed opaque, as is found with crystals grown from pure solutions.[*] A lower bound to the uptake occurs at $2 \cdot 10^{-3}\%$, because the crystals were colorless at lower concentrations.

The second group is made up of those crystals formed at dye contents of $(1-4) \cdot 10^{-2}\%$; they have a cubic habit. Crystals grown at $\sim 2 \cdot 10^{-2}\%$ were strongly and fairly evenly colored, which is a sign of maximal uptake. Those grown at $3 \cdot 10^{-2}$ and $4 \cdot 10^{-2}\%$ were less strongly colored in their peripheral zones.

---

[*] Lead nitrate falls in class 3/2 of the cubic system, and pure solutions normally give opauqe milk-white tetrahedral cubes.

The Effects of Thionin Blue Concentration in the Solution on the Morphology of $Pb(NO_3)_2$ Crystals

Dye concentration in initial aqueous solution, %	Group	Habit	Transparency and color
$2.5 \cdot 10^{-3}$ $3.0 \cdot 10^{-3}$ $5.0 \cdot 10^{-3}$	I	Tetrahedral cube, with cube faces predominating	Selective uptake by cube faces. Cube growth pyramids blue-violet and transparent. Tetrahedron growth pyramids opaque and milk-white
$1.0 \cdot 10^{-2}$ $2.0 \cdot 10^{-2}$	II	Cube	Strongly and evenly violet
$3.0 \cdot 10^{-2}$ $4.0 \cdot 10^{-2}$	II	Cube	Blue-violet. Central areas darker than peripheral
$5.0 \cdot 10^{-2}$ $1.0 \cdot 10^{-1}$	III	Tetrahedral cube	Selective uptake by cube faces. Cube growth pyramids colored. Tetrahedron growth pyramids transparent and colorless
$2.0 \cdot 10^{-1}$ $3.0 \cdot 10^{-1}$ $5.0 \cdot 10^{-1}$	IV	Cube, tetrahedral cube, pentagondodecahedron	Central areas colored. Color mainly violet. Peripheral areas colorless, or nearly so; transparent

Note. The dye concentrations are for the aqueous solutions before the lead nitrate was added.

The third group is made up of those crystals formed at dye contents of $5 \cdot 10^{-2}$ to $1 \cdot 10^{-1}$ %; these have a cubic tetrahedron habit, and have transparent tetrahedron growth pyramids.

The fourth group includes those crystals in which the cubic form predominates, and which sometimes show tetrahedron and pentagondodecahedron faces. These have peripheral zones that are colorless, or nearly so. The coloring at the center and at the edge varies from case to case, and from light sky blue to blue-violet. These crystals formed at dye contents of $(2-5) \cdot 10^{-1}$ %; they were usually coated in amorphous films of excess dye. Some crystals grown at $(3-5) \cdot 10^{-1}$ % showed outgrowths of dye crystals, however; these outgrowths were possibly epitactic in type.

Crystals Grown from Bulk Solution

Crystals grown from bulk solution by slow evaporation did not differ much from those grown on slides from drops. They were exactly like the microcrystals in color and in balance between simple forms when the dye content of the solution was about $5 \cdot 10^{-2}$ %. At $(1-2) \cdot 10^{-1}$ % they had larger colorless areas than the microcrystals, and (in some cases) good pentagondodecahedron faces. Similar crystals formed at $(3-5) \cdot 10^{-1}$ %. Here, however, the crystals were usually strongly colored and had dye crystals upon them.

The data show that the dye concentration has large effects on the habit, on the transparency, and on the nature of the coloring. The effects resemble those of methylene blue, but the concentrations at which the effects are greatest are nearly ten times larger.

It is clear that the concentration of dye in the mother liquor determines some features of the crystals, other things being equal.

The effects are directly dependent on the adsorption mechanisms that operate at the various concentrations; they will be dealt with in detail in a future paper.

SUMMARY

1. $Pb(NO_3)_2$ crystals have morphologic features that depend on the thionin blue concentration used; these features have been used to divide the crystals into four groups.

2. The dye is not absorbed at concentrations below $2 \cdot 10^{-3}\%$.

3. There is a maximum uptake. The even and strong color, and the cubic habit, indicate that the maximum occurs at about $2 \cdot 10^{-2}\%$.

4. The color in the crystals is weaker at concentrations of $(1-3) \cdot 10^{-1}\%$; the peripheral areas may be colorless.

I wish to thank N. G. Martynova, who helped me in this work.

LITERATURE CITED

[1] N. E. Vedeneeva and E. N. Slavnova, Trudy Inst. Krist. Akad. Nauk SSSR No. 7, 135-158 (1952).

[2] P. P. Orlov, Mosk. Obshch. Ispyt. Prirody No. 4, 1-60 (1896).

[3] P. A. Zemyatchenskii, Zap. Imp. Akad. Nauk St. Petersburg 1, Ser. 8, Fiz.-Mat. Otdel. 30, 30, 1-19 (1911).

[4] F. Ehrlich, Z. anorg. u. allgem. Chem. 203, 26-38 (1932).

[5] C. Frondel, Amer. Miner. 25, 91 (1940).

[6] P. Gaubert, Compt. rend. 180, 378 (1925).

[7] H. Buckley and H. Cocker, Z. Krist. 85, 58-75 (1933).

[8] W. France, Coll. Chemistry 5, 443-457 (1944).

[9] C. Bunn and H. Emmett, Disc. Faraday Soc. 5, 119-132 (1949).

[10] J. Whetstone, Disc. Faraday Soc. 16, 132-140 (1954).

[11] J. Whetstone, Trans. Faraday Soc. 51, 8, 1142-1153 (1955).

[12] J. Whetstone, J. Chem. Soc. 4841-4847 (1956).

[13] J. Whetstone, J. Chem. Soc. 4284-4294 (1957).

[14] E. N. Slavnova, Doklady Akad. Nauk SSSR 106, 6, 1007-1010 (1956).

[15] E. N. Slavnova, Doklady Akad. Nauk SSSR 107, 5, 693-696 (1956).

[16] E. N. Slavnova, Trudy Inst. Krist. Akad. Nauk SSSR No. 12, 98-110 (1956).

[17] E. N. Slavnova, Growth of Crystals, I [in Russian] (Izd. AN SSSR, 1957).

[18] P. Gaubert, Compt. rend. 190, 1230 (1930).

[19] H. Buckley, Z. Kryst. 76, 147 (1930).

[20] W. France, Reprint from Coll. Symposium Annual, 60-87 (1930).

[21] A. Neuhaus, Z. Kryst. 103, 5, 297-326 (1941).

[22] V. G. Khlopin and M. A. Tolstaya, Zh. Fiz. Khim. 14, 7, 941-952 (1940).

[23] E. M. Ioffe and V. L. Nikitin, Izv. Akad. Nauk SSSR, Otdel. Khim. Nauk No. 1, 15-22 (1943); No. 3, 191-7 (1943).

[24] G. Blisnakov and E. Kirkova, Z. Phys. Chem. 206, 271-280 (1957).

[24a] G. Blisnakov and E. Kirkova, Zh. Neorg. Khim. 3, 2, 517-525 (1958).

[25] H. Buckley, Crystal Growth [Russian translation] (IL, 1954) p. 325.

THE GROWTH FORMS OF THIAZINE DYE CRYSTALS

N. M. Melankholin

The absorption spectra and other optical parameters of organic dye crystals have been studied for several years in the Crystal Optics Laboratory of the Institute of Crystallography, Academy of Sciences, USSR; the results have in part been published [1]. Many growth forms have been examined, in order to find forms with optical orientations suitable for our purposes. Some of the growth forms we found have not previously been described, while others are interesting cases of forms already known. The data are of interest in morphology and growth form contexts. This paper deals with the growth forms of the thiazine and oxazine dyes we have studied.

The crystals were always made by letting drops of solution evaporate slowly on glass surfaces. A very little HCl was sometimes added to the solution, and usually assisted the growth. The HCl could not produce any fresh colored compounds, because all the dyes were present as chlorides. The hydrogen ions sometimes produced the leuco-forms [3], whose crystals were often seen. These crystals are colorless, though, and so are easily distinguished from the others.

Not all dyes show a large variety of growth forms. Some, such as methylene blue and thionin violet, give crystals of needle habit only, although at first sight characteristic rosette forms are seen along with crystals of good shape.

Methylene blue also gives crystalline films, i. e., thin crystalline layers which are continuous in isolated areas. These films are sharply bounded spots of rounded or irregular shape, and of blue, rose or violet color. Examined between crossed nicols, these films are seen to be each composed of several areas that extinguish almost simultaneously. The dichroism and conoscopic figures given by the films and by oriented crystals indicate that the colors correspond to differing orientations. Films of differing colors often come in contact, e. g., a blue film encounters a rose zone; there is then always a sharp boundary between the colors. These films are mosaic platelet crystals, which are about $200\,\mu$ across, but only 0.2-$0.3\,\mu$ thick.

Novomethylene blue and methylene green show the greatest variety of forms, especially if a minute trace of HCl is present. The former dye rarely gives regular crystals. The edges of the specimen, i. e., areas in which crystallization starts, sometimes show regular crystals as small rectangular or rhombic plates (Fig. 1), which shapes correspond to two different orientations. The monocrystals are sometimes more complex in shape, and are perhaps parallel intergrowths or filled-in skeletons (Fig. 2); sometimes they are typical dendrites. These crystals draw off much of the solution, and deplete nearby areas to do so. The glass is coated with very small crystals outside such areas.

At the start of evaporation we sometimes find bundles of fairly large needles crystals, but most of the crystals given by novomethylene blue have no clear crystallographic outlines. This is so for all crystals, including the rhombs and rectangles*noted above.

Thin plates of roughly elliptic form are the most common. These have small holes and dentate edges (Fig. 3). These plates are rarely monocrystals; usually they are mosaics, or intergrowths with sectors having differing orientations. Sometimes, especially if they are large, they pass into thin-branched dendrites, which branch randomly on all sides, a thing that normal dendrites do not do (Fig. 4a). These rosettes are sometimes as much as 1 mm in diameter. Large round holes are common in such rosettes; the holes contain thin outgrowths from the dendrites (Fig. 4b). Normal skeletal crystals are sometimes also found.

---

* The orientations were determined from the conoscopic figures, and were checked from the pleochroism.

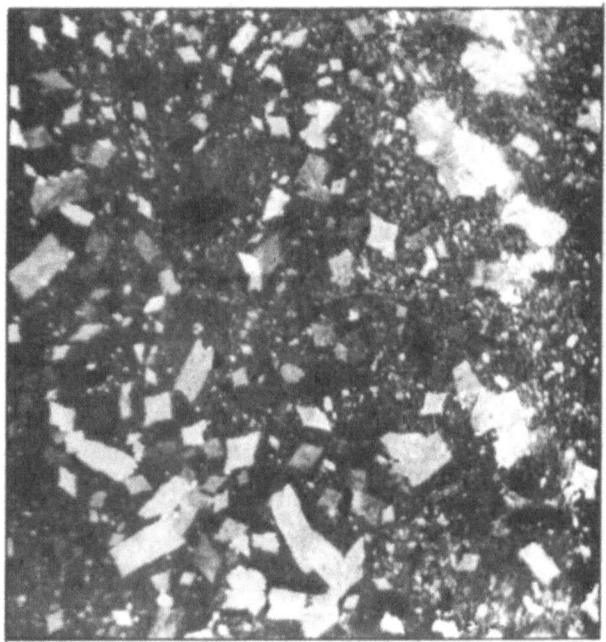

Fig. 1. Regular crystals of novomethylene blue.

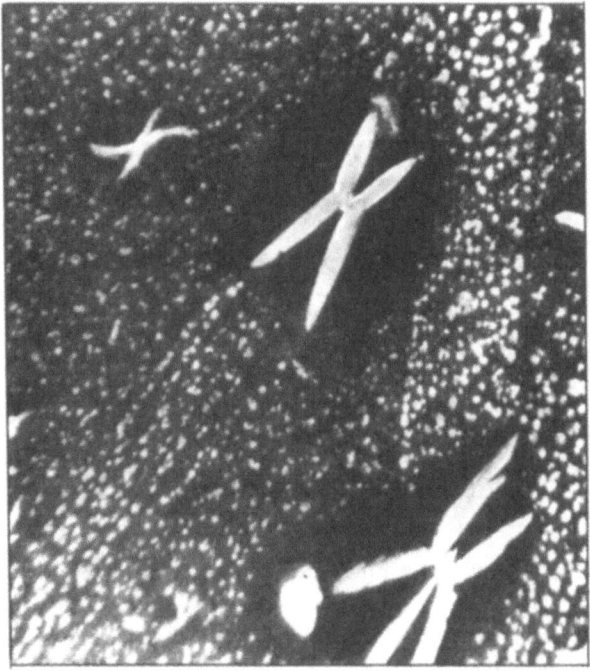

Fig. 2. Novomethylene blue. Large monocrystals.

Highly branched forms reminiscent of dendrites are also found; an especially interesting one is the skeletal "two-sheet" form shown in Fig. 5. Platelet crystals suually grow as rosettes, but very strange forms that resemble flowers are on occasion seen, especially when platelet and needle crystals combine.

This variety in the growth forms of novomethylene blue (and of methylene blue) might indicate that the dye exists in different modifications, or that traces of some other similar substance are present. However, the

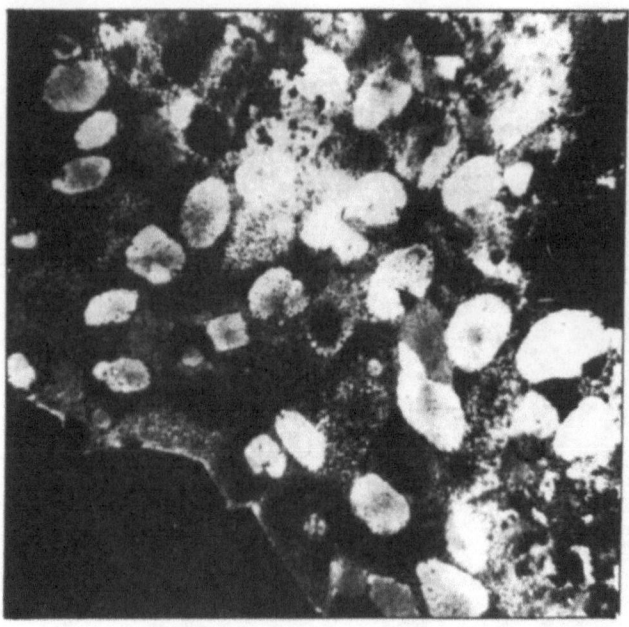

Fig. 3. Novomethylene blue. Rounded platelet crystals.

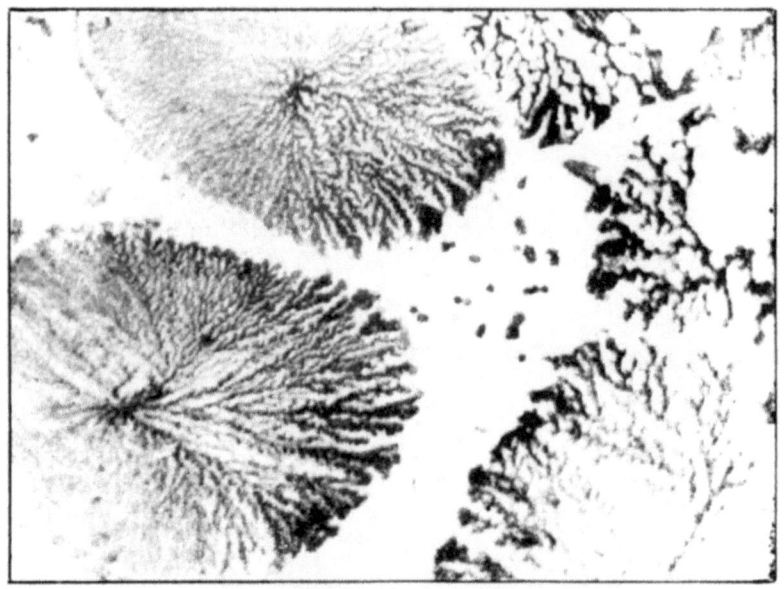

Fig. 4a. Novomethylene blue. Randomly branched dendrites.

absorption curves and the pleochroism show that all these crystals are the same, except that they are differently oriented. The orientation may be dependent on the state of the glass; the molecules that go to form the nuclei may be differently oriented as a result.

The growth conditions (supersaturation, layer thickness, evaporation rate) also change during the growth. The crystals formed while the layer is thick find conditions that differ from those that apply to crystals that grow later, i. e., when the layer is thinner and is evaporating more rapidly. These changes may alter the habit in different ways even for crystals that starting growing at the same time. These effects cannot occur when large volumes of solution are used, because the growth conditions scarcely change.

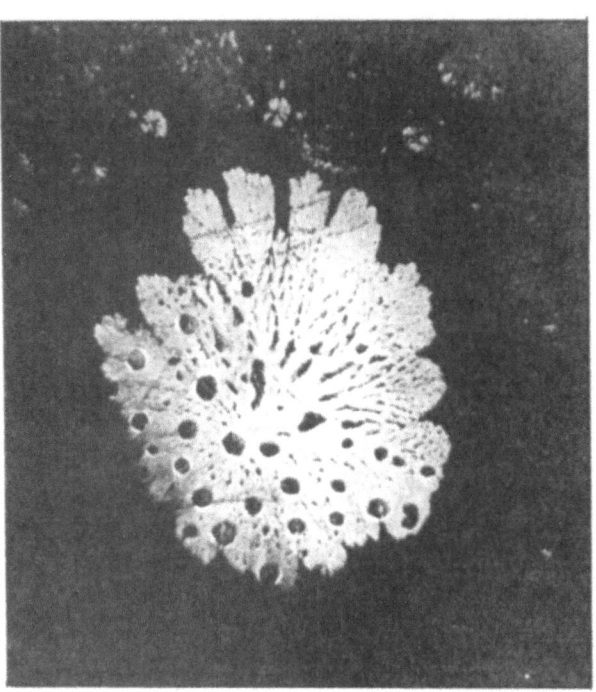

Fig. 4b. Novomethylene blue. Dendritic growth with holes.

Methylene green shows no less variety in its growth forms. We find rhombic, rectangular, square and curved-needle crystals, branched intergrowths, divergent bundles, shapeless plates, rosettes, and rose, blue, and zoned spherulites. The needle crystals give several types of intergrowth, especially branched ones. Some crystals in such forms are straight, but most are curved. Splitting is usually responsible for the curvature; though thick unsplit needle crystals may also be bent (Fig. 6); a single intergrowth may often show straight and bent crystals.

The spherulites (especially rose-colored ones) are the most typical forms, and next to them the very similar radial intergrowths. These latter are often made up of compact bundles, which diverge and produce forms resembling crystals composed of two or more leaves.

The rose and blue spherulites differ both in color and in structure. The rose ones are almost always of good shape, with very thin component crystals (sometimes too thin to be distinguished). The blue ones have coarser structures, and the component fibers often cross. The two types will grow in contact with one another (Fig. 7). The blue grow faster, and may engulf the rose. It is clear that we do not have two modifications of the dye (although

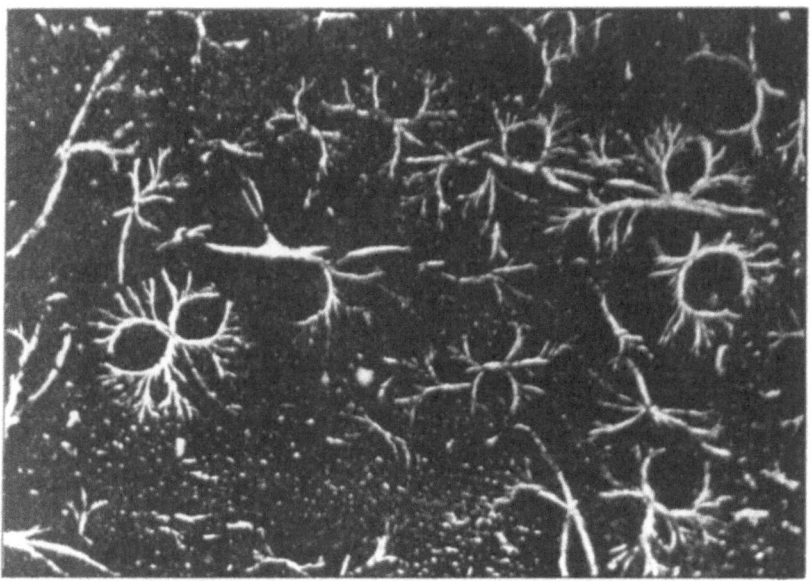

Fig. 5. Novomethylene blue. Branched sets of needle crystals.

we might have expected this from the form of the spherulites), but merely two different orientations relative to the glass. The fact that the two types have the same absorption bands points to the same conclusions.

The rose spherulites are sometimes surrounded by halos of grey or blue minute crystals, which have a different orientation. These latter have grown under different conditions, and the change in the conditions has presumably caused the spherulites to stop growing. The rose ones are often zoned (with concentric rings); the rings have two different colors that alternate when viewed between crossed nicols (Fig. 8). We may conclude from this that the component fibers are twisted, and thus show periodically varying birefringence.

Fig. 6. Methylene green. Bent prismatic crystals.

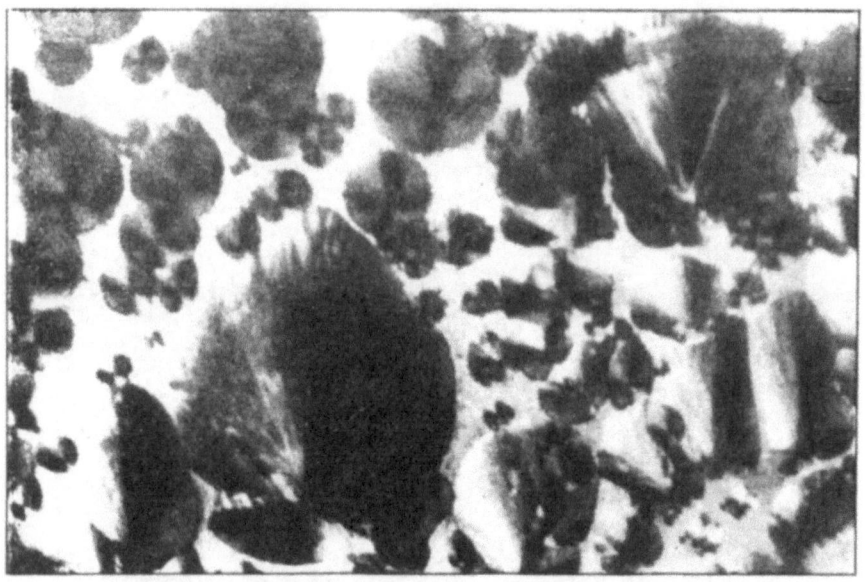

Fig. 7. Methylene green. Rose and blue spherulites.

The dichroic colors of the zones correspond to the colors found with long prismatic and rhombic crystals in such preparations. From this fact we can derive the optical orientations of the component fibers, which lie along the N_m (β) axis of the indicatrix.

The zones are sometimes all of the same width, but they normally decrease gradually in width as growth proceeds (Fig. 8). The fibers twist more rapidly because the layer of solution is thinner and the growth conditions have altered.

This zone structure reveals very well the splitting or branching that occurs during growth in these spherulites; the cause would appear to be that there is not enough solution for the spherulite to grow all round. A spherulite

Fig. 8. Methylene green. Zoned spherulite consisting of twisted crystal fibers.

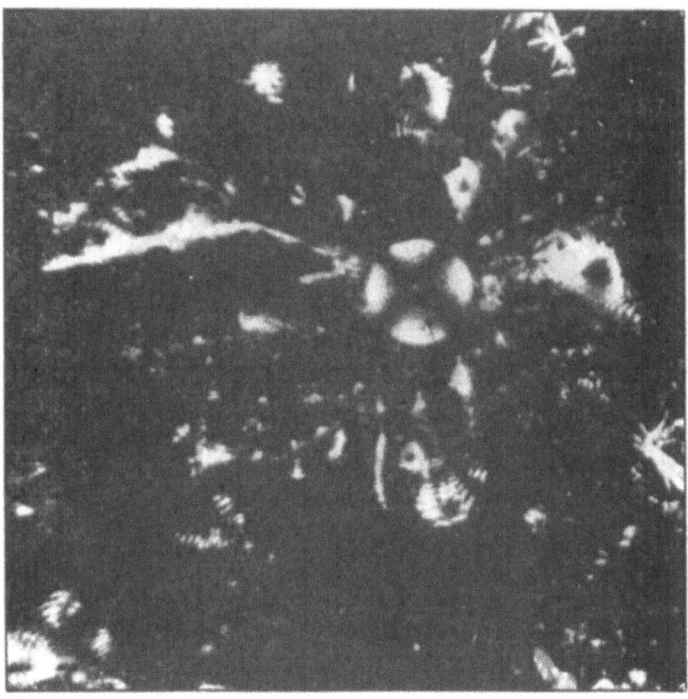

Fig. 9. Methylene green. Branched zoned spherulite grown in an inadequate amount of solution.

that has reached a certain size ceases to grow at all points, and continues to grow only at a few isolated places, which act in effect as the centers of fresh spherulites that grow as round sectors, which may themselves branch (Fig. 9).

Fig. 10. Crystals of two different modifications of the lecuobase of methylene green.

Fig. 11. Thionin blue. Skeletal crystals and intergrowths of branched crystals.

One special growth form is quite common in these methylene green preparations. In many cases some areas of the preparation are covered with small rounded drops, which are birefrigent and dichroic. Viewed between crossed nicols, these drops appear as something between compressed amorphous lumps and deformed crystals; they are oriented over quite large areas of the glass. Their dichroism and birefringence increase with time. It may be that these are drops of amorphous dye with their molecules oriented in more or less the same way; the degree of order gradually increases, because the polished glass has the power to orient the molecules.

Fig. 12. Thionin blue. Twisted radial growths.

Fig. 13. Needles of capri blue grown in a space free from solution.

Colorless crystals of the leucobase can also be seen on occasion. These are of two types, which presumably belong to two different modifications. One gives small highly birefringent X-shaped crystals (Fig. 10), which are polysynthetic twins. The other (which is more common) gives larger weakly birefrigent rectangular platelets, which have the peculiar feature that on the top of each there is a square outgrowth with rounded corners, which is bisected by two lines parallel to the sides of the square (Fig. 10).* The cause of these outgrowths is not clear.

---

* Oblique illumination shows that they are in fact outgrowths, and not pits.

Thionin blue usually gives an amorphous film, from which crystals grow slowly. Skeletal crystals and radial patterns result. The skeletal crystals are long rectangles, with branches of the dendrite parallel to the diagonals (Fig. 11). The radial growths consist of long crystals that widen or branch at the ends. These crystals often tend to bend, usually in the same direction for all crystals within a group.

The result is to produce spiral or twisted patterns (Fig. 12). The crystals are very thin and fiber-like. Thionin blue crystals usually bend during their growth. Even broad crystals may bend if the growth conditions are appropriate. Small crystals are boat-shaped, but become skeletal as they grow. Curved and straight crystals often grow together (Fig. 11).

We also examined one oxazine dye, namely capri blue (the others were closely related thiazine dyes). The growth forms are very like those of thionin blue. The solution dries up to leave a dark-blue amorphous mass, in which long crystals grow slowly; these crystals soon become dendrites, which may be very complicated. A special type of growth not found with the other dyes occurs here. If the crystals are growing rapidly, it sometimes happens that wide gaps free from amorphous material form between the crystals during the early stages. Then needle- or knife-shaped outgrowths appear in these gaps, and may become very long (Fig. 13). These crystals grow in areas free from dye, and so they grow at the expense of scraps of residual amorphous material left between the preexisting dendrites. They grow mainly on the faces at their ends, so the material must reach the ends by migrating along the long faces of the crystals [2].

LITERATURE CITED

[1] I. M. Melankholin, Optics and Spectroscopy (USSR) 3, 2, 104-114 (1957).

[2] A. V. Shubnikov, Formation of Crystals [in Russian] (Izd. AN SSSR, 1947).

[3] I. M. Kogan, The Chemistry of Dyes [in Russian] (Goskhimizdat, 1956).

THE GREEN AND BROWN COLORS OF SYNTHETIC
QUARTZ CRYSTALS

L. I. Tsinober, L. G. Chentsova, and A. A. Shternberg

<u>I.</u> Green and brown crystals have been produced from aqueous solutions of K_2CO_3 under laboratory conditions. In nearly all cases the steel autoclave was found to have corroded greatly, and to be coated with a greenish material. The seeds were slices cut parallel to the planes (0001), ($\overline{1}011$) or <u>r</u>, ($10\overline{1}1$) or R, or were slices taken at an angle to the R plane (Fig. 1).

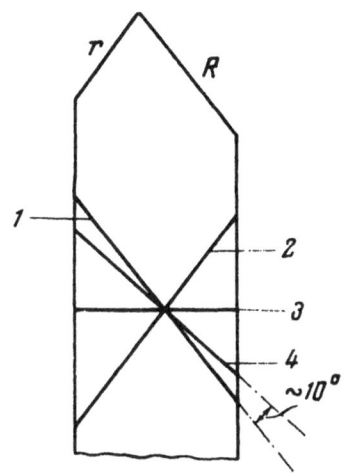

Fig. 1. Orientations of the plates used as seeds: 1) ($10\overline{1}1$); 2) ($\overline{1}011$); 3) (0001); 4) oblique irrational section at an angle to the R plane.

A series of tests at various initial K_2CO_3 concentrations revealed the more notable features of the process that produces colored crystals. At a concentration of a few percent of K_2CO_3 the crystals were mainly brown. The depth of the color was related to the time the autoclave had been working. In a series of eight tests in one autoclave we found that the color appeared in the third test. The shade was deepest in the sixth, and thereafter lightened.

The color deepened as the K_2CO_3 concentration was increased, and eventually appeared in the first test (with an unused autoclave). The color became green when the concentration was very high, though the brown color persisted in isolated spots in the crystals.

<u>II.</u> The crystals showed sector and zone distributions of color. Polished sections cut parallel to ($11\overline{2}0$) usually showed the deepest color associated with the <0001> growth pyramid; at low alkalinities this pyramid was mainly brown, and at high ones green. At low and medium alkalinities, when the main body of the <0001> pyramid was brown, a layer 2-4 mm thick around the seed was often green. The green passed gradually into the brown. The brown color appears because the alkalinity falls as the heavy phase [1, 2] is formed.

The growth pyramids on the section inclined to R were always brown, no matter what the growth conditions. The color was often as deep as that in the <0001> pyramid. Now the pinacoid growth surface was often highly developed (especially with the cross-cut seeds), and a second-order growth pyramid was formed in the <0001> pyramid in sections cut at an angle to R, so it is clear why brown bands occur in the green <0001> pyramid, and also why brown wedge-shaped growth pyramids form on the ($11\overline{2}0$) and ($\overline{1}011$) faces (Fig. 2).

The growth pyramids on the <$11\overline{2}0$> and <$\overline{11}20$> trigonal prisms, and that on the very slow-growing <$10\overline{1}0$> hexagonal prism, were usually of a weak green color. The pyramids on <$\overline{1}011$> and <$10\overline{1}0$> (Figs. 4 and 5) were never colored if the growth was normal.

These latter pyramids could become colored, however. If the growth rate increased suddenly (which can be detected from the curvature of the bounding surface between the pyramids on the rhombohedra and pinacoid), some parts of the pyramids on the rhombohedra could become a weak grey-green color (Fig. 2).

This dependence of color on growth rate is best seen with the <0001> pyramid, whose growth rate can be varied greatly by adjusting the supersaturation. If the rate for <0001> was reduced to 0.3-0.4 mm/day (for both

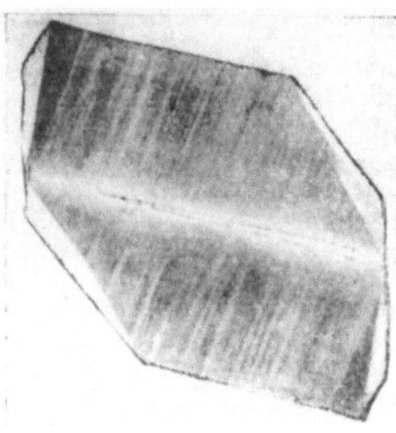

Fig. 2. Plate cut parallel to (11$\bar{2}$0) from a crystal grown from a seed cut at 12° to <0001>. × 1.

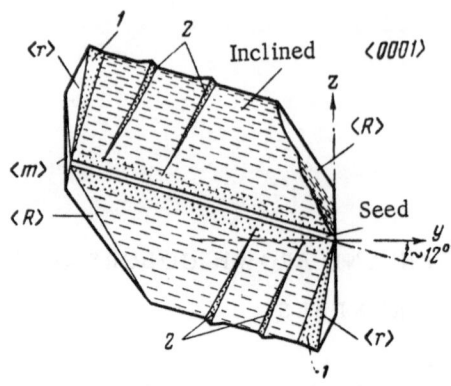

Fig. 3. Explanation of Fig. 2: 1) wedge-shaped brown pyramids on ($\bar{1}$011) and (0001) faces; 2) brown tails, which are second-order growth pyramids (inclined to R) in the inclined <0001> pyramid, which is green. The hatched areas are green, the dotted ones brown.

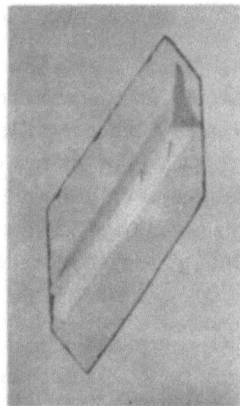

Fig. 4. Photograph of a plate from a crystal grown from a seed cut parallel to a face of the <$\bar{1}$011> negative rhombohedron. × 1.

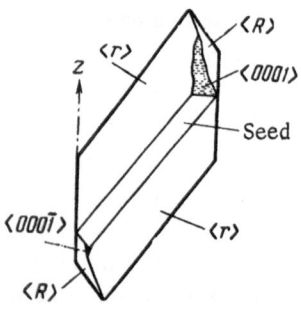

Fig. 5. Explanation of Fig. 4. The notation is the same as in Fig. 3.

sides of the seed), the color became much weaker, even if the K_2CO_3 concentration was very high. Completely colorless crystals can be made if the K_2CO_3 concentration is reduced and the growth rate is kept at this low value.[*]

"Sector polychrome" and "zone polychrome" are two adjectives that may well be used to describe these features of the coloring.

Sector polychrome refers to the coloring that occurs when the simultaneously grown growth pyramids of the various simple forms have different colors. The effect is controlled by the structural and energy relations between the lattice planes of the various faces, by the structure of the adsorbed impurity, and perhaps by metrical relations between the planes also. An example is the green color formed in the <0001> and <10$\bar{1}$1>[**] pyramid, with brown in the pyramid on the plane inclined to R (denoted by <o. t. R >), and no color at all in the <10$\bar{1}$1>[**] and <$\bar{1}$011> pyramids (Figs. 6 and 7).

Zone polychrome refers to the coloring that occurs when parts of the same pyramid that have grown at different times have different colors. This effect is caused by changes in the form in which the impurity enters the crystal, i. e., ultimately by changes in the physicochemical parameters. The brown color that replaces the green in the <0001> pyramid is an example.

There are also minor zonal changes in the strength of the color (this applies mainly to the green). The effect is to be considered as caused by changes in the normal growth states, or in surface regeneration, and by changes in the concentration of the impurity.

---

[*] It may also be that the temperature and pressure themselves directly affect the coloring, in addition to any effect caused via the growth rate and alkalinity. This topic is not touched on here because it has not been examined.

[**] As in original — Publisher's note.

Fig. 6. A polished plate cut from a crystal grown from a crystal cut at an angle to R. × 1.

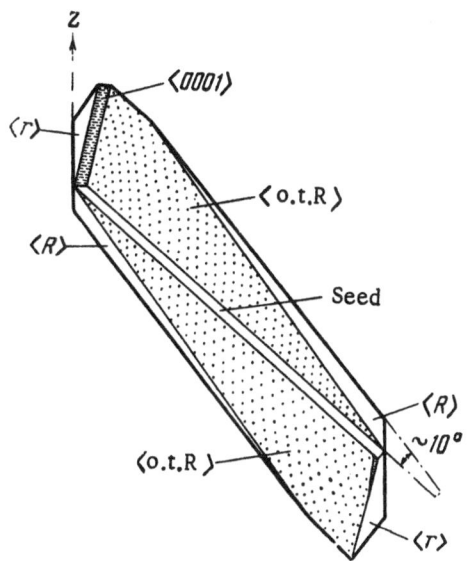

Fig. 7. Explanation of Fig. 6. The notation is the same as in Fig. 3.

III. Spectral analysis * showed that the colored crystals contained more iron that did those grown under normal conditions; the color may be assumed to be caused by iron ions. To confirm this we recorded the absorption spectra of the colored crystals, and the changes produced by a dc electric field applied at a high temperature.

We used an SF-4 spectrophotometer fitted with a polarizer to record the optical density in the 250-1200 $m\mu$ range. Plates were cut parallel to c from the < 0001> pyramid for this purpose. The area used (0.5 cm^2) usually covered several zones. All samples, whether green or brown, also showed brown streaks along c. The results are thus average values. Figure 8 shows the most typical curves for green (1_o and 1_e) and brown (2_o and 2_e) samples. The wavelength is given in $m\mu$, and the optical density D as k = D/d, where d is the thickness of the plate.

The green samples show maxima at 740 and 900 $m\mu$, and little absorption at 400-600 $m\mu$. The dichroism is small, increases towards short wavelengths, and changes sign at about 580-600 $m\mu$; the ordinary ray is the more strongly absorbed one in the ultraviolet and in most of the visible region, whereas the extraordinary is the more strongly absorbed in the infrared.

The brown samples absorbed much more strongly than the green ones in the ultraviolet. The absorption dropped in the visible, and was almost constant above 700-750 $m\mu$. The dichroism was large in the ultraviolet and over much of the visible regions, and showed the same general trend as did that for the green samples. It may be that the brown streaks are responsible for the dichroism in the green crystals. Some of the brown samples showed weak peaks at 740 and 900 $m\mu$, which shows that centers of the type responsible for the green color were present.

These curves were compared with the curves for several minerals whose colors are caused by iron. Those for the brown crystals were very like those found by Melankholin [3] for hematite (Fe_2O_3). Those for the green were like those found for mica colored by Fe^{2+} ions [4]. The maxima are at 740 and 900 $m\mu$ in both cases. The quartz absorbs less strongly than the mica in the visible and ultraviolet. Micas contain much Fe^{3+} as well as Fe^{2+}, and so the absorption rises rapidly at short wavelengths.

IV. The colors were not affected by heating the samples to 500C°. The brown color became green, because Fe^{3+} was reduced to Fe^{2+}, when a plate was heated to 400-500°C for several hours with a direct current passing through it. If this current passed at right angles to c, the Fe^{2+} was oxidized back to Fe^{3+} when the plate was again heated to 500°C; the brown color was thereby restored. This effect did not occur if the current flowed along c. The Fe^{3+} ions, and the Fe^{2+} ions that result from them, take up positions in the lattice different from those taken up by Fe^{2+} trapped during growth. An electric field acting along the structural channels displaces the Fe^{2+} ions formed by reducing Fe^{3+} to positions at which they cannot be oxidized back to Fe^{3+}.

* G. D. Pakhomova did the analyses at VNIIP; the data show that the strongly colored pyramids contained 0.01-0.1% of iron, whereas colorless parts of the same crystals, and crystals grown from Na_2CO_3 solutions, contained 0.003-0.007% iron.

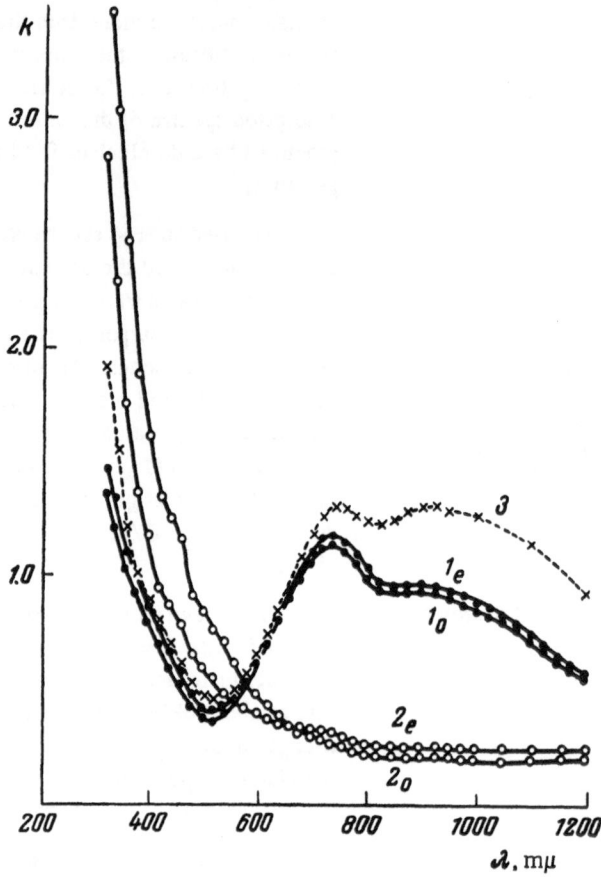

Fig. 8. Absorption curves for the ordinary and extraordinary rays ($\underline{o}$ and $\underline{e}$, respectively): 1_o and 1_e) green quartz; 2_o and 2_e) brown quartz; 3) quartz that became green when a current was passed through it.

The quartz with reduced iron ions showed no dichroism at all, and the spectra were very similar to those for green quartz. This result confirms our assumption that the dichroism in green crystals might be due to the brown streaks.

The dichroism in brown quartz must occur as a result of some anisotropic interaction between the Fe^{3+} ions and the quartz proper. The Fe^{2+} ions show no such interaction.

The green color lightens if the current is passed along $\underline{c}$ for a long time. The effect occurs most rapidly at the anode end, but thin colorless layers, such as are found in smoky quartz [5], do not form. The color boundary did not move in crystals having colorless zones near the cathode.

This loss of color can be explained in terms of ionic migration along the structural channels. The loss occurs gradually throughout the volume, and so the ions responsible for the color do not accumulate in the colorless zone.

$\underline{V.}$ These data, and those on electrothermally treated colored quartz, enable us to draw some tentative conclusions on the nature of the effect.

1. Aqueous solutions of K_2CO_3 attack steel autoclaves at high temperatures and pressures, and the solutions become rich in iron.

2. The green and brown colors are caused by the crystals' absorbing Fe^{2+} and Fe^{3+} ions, respectively. Both colors show characteristic sector and zone distributions.

3. The ratio of divalent to trivalent iron in the solution increases with the K_2CO_3 concentration, and so favors a green color in the crystals.

LITERATURE CITED

[1] V. P. Butuzov and L. V. Bryatov, Kristallografiya 2, 5, 670-5 (1957).

[2] S. Taki, Naturwissenschaften 44, 23, 614 (1957).

[3] N. M. Melankholin, Zap. Vses. Min. Obshch. 75, 2, 89-93 (1946).

[4] S. V. Grum-Grzhimailo, L. I. Anikina, E. N. Belova, and K. I. Tolstikhina, Mineral. Sborn. Lvov Geol. Obshch. 9, 90-119 (1955).

[5] L. G. Chentsova and N. E. Vedeneeva, Doklady Akad. Nauk SSSR 68, 305-8 (1949).

THE CRYSTALLIZATION OF KI IN CONTACT WITH
BIOTITE AND MUSCOVITE

I. E. Kamentsev

Deo et al. [1] have recently studied how oriented crystals are formed on calcite and mica. They found that oriented crystals that formed on one surface continued to grow with the same orientation even if they encountered another surface with a different orientation. However, they concluded that in epitactic growth the surface determined the orientation of the initial nucleus, but that the subsequent growth was not influenced; this contradicts certain experimental data [3].

We have studied this effect with KI crystallizing at the natural contact between two micas (biotite and muscovite), which have different lattice parameters. Only crystals that encountered such contacts during growth were examined.

Crystals of KI grow on the (001) faces of biotite and muscovite from their octahedron faces [7], because the planes in contact have similar structures; (001) in the mica is parallel to (111) in KI, and the rays in pressure stars in the mica are parallel to ($0\overline{11}$) in the KI, so the three-fold axis in the KI is normal to the (001) plane in the mica.

The octahedral nets in KI differ from the parameter of (001) in mica by 3.8% for muscovite, and by 6.2% for biotite. Nuclei appear more rapidly on muscovite than they do on biotite, and the nucleation probability is larger for muscovite [3].

The KI was allowed to crystallize from drops of saturated solution on fresh mica surfaces at room temperature. Crystals that grew from biotite to muscovite continued to grow unchanged. Those that grew in the reverse were usually truncated at the vertex which continued to grow on the biotite (Fig. 1).

This effect is not caused by steps at the contact between the biotite and muscovite, because the truncated vertex was always on the biotite, and not at the contact line between the micas. The effect is also not caused by slight differences in orientation between the two micas, because the growth from biotite to muscovite was normal. Numerous experiments gave us 63 KI crystals that had grown from muscovite to biotite, and of these 56 (some 90%) had one vertex truncated.

Fig. 1. Oriented KI crystals growing at a muscovite-biotite contact. Truncated vertices are seen on the biotite.

Thus KI crystals growing from muscovite to biotite have the vertex which grows onto the biotite truncated.

DISCUSSION

Let us consider the growth mechanism of a KI crystal at such a contact. A definite supersaturation is required for an oriented crystal to grow on the muscovite [5]; a monolayer of a certain size is formed [2]. The growth starts on the muscovite but transfers to the biotite, which material demands a higher supersaturation before nuclei form [4]. This may explain why the vertex stops growing. The substrate influences the crystal growing on it, and reduces the supersaturation needed for the crystal to grow in a direction parallel to the oriented substrate.

This effect probably occurs in many other instances. Royer [8] has grown $NaNO_3$ as plates between mica crystals. The growth parallel to the surfaces no doubt occurs because the mica influences the crystallization.

Fersman [6] gives some interesting data on intergrowths of tourmaline with feldspar pinacoids.

He found that the faces of the tourmaline in contact with the feldspar pinacoid were greatly increased in area. The habit is affected because the tourmaline grows more rapidly parallel to the pinacoid; the feldspar reduces the supersaturation needed for the tourmaline to grow parallel to it.

LITERATURE CITED

[1] A. Deo, G. Finch, and F. Charpurey, Proc. Royal Soc. A236, 7, 1204 (1956).

[2] H. van der Merve, New Investigations in Crystallography and Crystallochemistry [Russian translation] (1950).

[3] P. A. Zemyatchenskii, Izv. Imp. Akad. Nauk 541 (1914).

[4] I. E. Kamentsev, Kristallografiya 1, 2, 240 (1956).

[5] D. E. Ovsienko, Problems of Physics of Metals and Metal Science [in Russian] (1953), No. 4.

[6] A. E. Fersman, Selected Works [in Russian] vol. 1, p. 51.

[7] K. P. Yanulov, Doklady Akad. Nauk SSSR 62, 813 (1948).

[8] M. Royer, Bull. Soc. France Min. 77, 1004 (1954).

Department of Crystallization, Zhdanov State University, Leningrad

DETERMINATION OF SURFACE ENERGIES OF CRYSTALS
FROM EQUILIBRIUM STATES

A. V. Belyustin

There is as yet no unanimity on the relation between surface energy and growth rate of a crystal face. It may be that, if the relation between surface energy and habit could be established, the measurement or calculation of surface energies would be facilitated for faces in contact with solution, melt, or vapor.

I have shown [1] that it may be possible to establish the surface free energy of faces from the equilibrium shape in a gravitational field. Previously I considered only a very simple cubic crystal. I found a formula [1], which may be replaced by one more convenient:

$$4 (a - h) \sigma_{12} = (\rho_1 - \rho_2) ah^2, \tag{1}$$

where σ_{12} is the surface free energy for a cube face in contact with a medium, h is height, a represents the transverse dimensions of the equilibrium crystal, ρ_1 is the density of the crystal, and ρ_2 is the density of the medium.

Let us now consider other relations, in which, as in (1), the surface energy may appear.

Firstly, if, when we seek the minimum in the free energy $F = V + W$ (where V is the potential energy and W is the total surface energy), we use the Lagrange multipliers method, instead of eliminating the variables as in [1], we get a formula of the type of (1) which is applicable to many crystals.

We can also use several systems, as in Fig. 1. The crystal is assumed cubic, for simplicity. Now (1) will apply to a free crystal at the bottom of a crystallizer (Fig. 1a), and so σ_{12} can be found, if equilibrium is reached exactly. Then, by using the crystal capillary (Fig. 1b), we can find σ_{14} also (the surface energy at an interface with air). Imagine a rectangular channel of sides a and b within the crystal, with the channel walls parallel to the cube faces. The liquid (e. g., saturated solution) within the channel will rise above the level outside. Now this part of the free energy must be a minimum, and it depends on the rise, so that

$$2 (a + b) (\sigma_{14} - \sigma_{12}) = \rho_2 abh. \tag{2}$$

Now σ_{12} and σ_{14} will also determine the equilibrium state of a crystal so placed on a fiber that only the lower part is surrounded by solution. The crystal will, say, extend downwards if material is only redistributed (Fig. 1c); the potential energy will decrease, whereas the surface energy will increase. Some state of local equilibrium (corresponding to a local minimum in F) will be reached. We assume that the surface energy will be increased greatly if a neck forms, and that the crystal therefore will not break. The equilibrium is then described by

$$(2h - a) \sigma_{12} + a (\sigma_{14} + \sigma_{24}) = \frac{1}{2} (\rho_1 - \rho_2) ah^2. \tag{3}$$

Here σ_{24} is the surface tension of the solution at an interface with air, and is a quantity that may be measured independently. Measurements with two different values of h and a give us two equations of the form of (3) for σ_{12} and σ_{14}. We can also combine (3) and (2). Thus we have available several ways of finding the quantities, and can choose the best, or compare results.

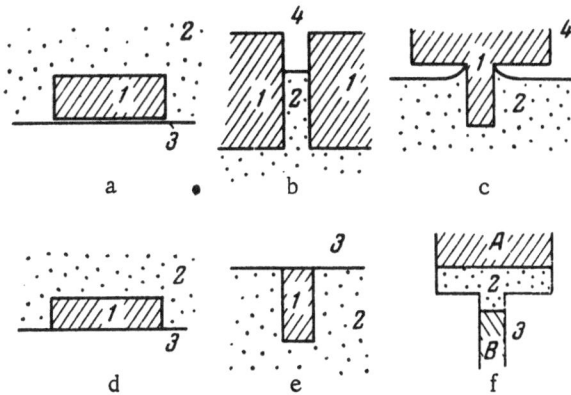

Fig. 1. Equilibrium systems: 1) crystal, 2) mother liquor,
3) solid, 4) gas.

In Fig. 1d (crystal on the bottom of the crystallizer) the solution for the minimum F is (7) of [1]. This we put in a form similar to (3):

$$(a - 2h)\, \sigma_{12} + a\, (\sigma_{23} - \sigma_{13}) = \frac{1}{2}\, (\rho_1 - \rho_2)\, ah^2. \qquad (4)$$

Then (4), applied to two crystals of different sizes, gives us two equations, in which the unknowns are σ_{12} and $\sigma_{23} - \sigma_{13}$, the difference between the values of the surface tension for the liquid-crystal and crystal-base interfaces. A crystal fixed to the lid of a full crystallizer (Fig. 1e) gives us a relation analogous to (4).

Finally, consider a combination of crystals in a tube (Fig. 1f). Suppose that the top (wide) part of the tube is square in cross section, with side l, and that the lower (narrow) part is rectangular, of sides $\underline{a}$ and $\underline{b}$ (other cross sections can be used). Crystal A lies at the top, crystal B at the bottom, with their horizontal faces cube faces. Between the crystals there is a layer of mother liquor. We assume that B is turned about a vertical axis relative to A. We also assume that A presents cube faces to the wall, whereas B presents {110} planes. Then the surface energies between crystal and wall (σ_{13} for A and σ'_{13} for B) will differ.

The potential energy will decrease as material is transferred from the top crystal to the lower one, but the surface energy increases if $\sigma'_{13} > \sigma_{23}$ (if the converse is true, the tube must have its narrow end at the top). This transfer changes the distance Δh between the faces of crystals A and B. The minimum free energy occurs when

$$(1 - \varkappa)\, \sigma_{23} + \varkappa \sigma'_{13} - \sigma_{13} = \frac{1}{4}\, l\, (\rho_1 - \rho_2)\, \Delta h. \qquad (5)$$

Now if $\varkappa = \dfrac{a + b}{2ab}\, l$ is large enough we can neglect σ_{13}. The tube need not be of glass; it can be made of a crystal that does not react with the solution. We can make σ_{13} much smaller than σ'_{13} and σ_{23} by choosing the material and orientation suitably. Measurements repeated with two different values of $\varkappa$ give us two equations for the two unknowns σ_{23} and σ'_{13}.

It is clear from the above discussion that these various equilibrium states give us several ways of determining the surface energy.

Only the experiment with the crystalline capillary is convenient for actual use. The transfer of material from one face to another is very slow, and long times are needed to reach equilibrium with medium-sized crystals. It is not, however, in principle impossible to approach the equilibrium state sufficiently closely for crystals on which proper measurements could be made.

LITERATURE CITED

[1] A. V. Belyustin, "The equilibrium shapes of crystals in a gravitational field," Kristallografiya 2, 5, 590 (1957).

Physicotechnical Research Institute, Gorky University

II. GROWING MONOCRYSTALS

(Apparatus, Methods, and Accessory Operations)

GROWING CALCITE AND OTHER CARBONATES*

Jan Kaspar

Monocrystals of calcite, and of the other alkaline earth carbonates, are of theoretical and practical interest. It would be of interest to see how the form of the crystal is related to the medium from which the crystal grew, since certain simple carbonates, especia ly calcite, show very many and varied forms. It is also necessary to establish the degree of isomorphism of the calcite-group minerals, and the positions of minerals such as monheimite, cobaltsmithsonite, oligonite, and so on. Calcite is of great value in optical instruments. It is therefore only to be expected that calcite and related carbonates should be studied closely.

The first experiments on synthesizing calcite were done 120 years ago, and there are now very many papers on the topic. The other carbonates have been studied more recently. Only a few preliminary tests have as yet been made in relation to cobaltocalcite (spherocobaltite) and otavite.

The data given in some short reviews, and by Hintze [4], Doelter [3], Mellor [6] and Dana [2] show that the experiments are of two types:

1) those in which the crystals are grown physically, especially by methods in which solubilities are used, and

2) those in which chemical methods are used, i.e., a) slow precipitation caused by running in a soluble alkali carbonate, and

b) double decomposition at high temperatures (up to 200°C) between a carbonate and a soluble salt (chloride, nitrate or sulfate) of the metal in question.

This division is only rough. Two methods are combined in some cases, e.g., when bicarbonates are used.

Physical methods are alone used to grow monocrystals at present. It may be that such methods are suitable for the calcite-group minerals, although these are very sparingly soluble. Ikornikova and Butuzov [5] have recently given some data on the method.

I have used the chemical methods. My results indicate that my method, and Ikornikova and Butuzov's, may both lead to the same result.

Rose's [7] is the oldest synthesis of calcite. Rose reacted ammonium carbonate with calcium chloride in 1837.

I also have used this method. It is, however, very difficult to use inorganic compounds for this purpose. I therefore tried an organic compound, namely urea.

Urea is formed when gaseous carbon dioxide reacts with gaseous ammonia:

$$CO_2 + 2NH_3 \rightarrow H_2O + CO(NH_2)_2.$$

The reaction is, under suitable conditions, reversible, and so urea can be made to give ammonium carbonate. The reaction $CO(NH_2) + 2H_2O \rightarrow (NH_4)_2CO_3$, if controllable (especially if the rate can be made low), could be used in growing carbonate monocrystals.

---

*Paper read at the Second All-Union Conference on Chemical Crystallography.

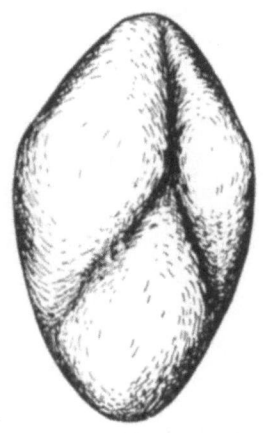

Fig. 1. Form of a calcite crystal at a late stage of growth.

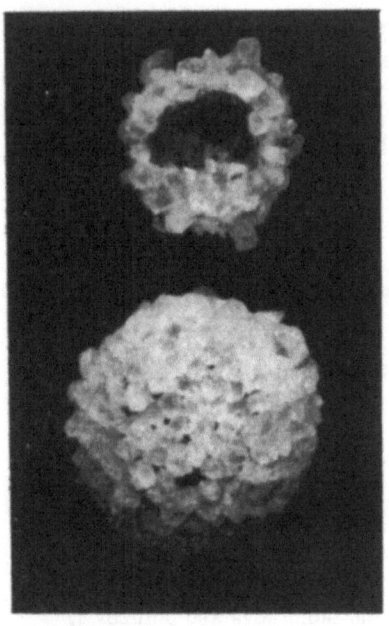

Fig. 2. Polycrystalline films of calcite on gas bubbles.

It is possible to decompose the urea catalytically at low temperatures, or with hot acids or alkalis. On Wichterle's advise I have used urease at low temperatures (up to about 50°C). The best temperature is about 35°C, but all the tests to be dealt with here were done at 20°C.

The urea was mixed in solution with a soluble calcium salt (acetate, chloride or nitrate). A little soya-bean urease was added to the solution.

The crystallization was observed in an open vessel, and in a wide sealed tube, with or without seeds. Small Iceland spar rhombs were used as seeds.

The urea solutions were of half-molar, molar, and twice molar strengths. The calcium salt solutions were in all cases molar. The solutions were mixed 1:1, and to 200 ml of the mixture was added 4 ml of urease solution. The reagents were all of chemically pure grade.

The pH of the mixture was 5.5 at first, but it soon rose to 6.5-7. The pH was measured with an indicator. Each test lasted 10 days.

I found that the calcite deposited at nearly equal rates in the two types of vessel. A seed roughly doubled the deposition rate. If the seed was large, its surface became coated with polycrystalline calcite. The reason was that the urease solution contained some protein impurities that affected the nuclei. Seeds of size a few tenths of a millimeter (fine calcite sand) gave satisfactory results.

The crystals had rounded faces, with accessory faces. They were not, however, simple rhombs, although others have found such forms.

A seed at first became coated with a fairly uniform layer. Later on it became clear that the form was that of a higher rhombohedron (probably 2R), as shown in Fig. 1 (this effect also occurred with crystals grown without seeds). These results were used in some later studies on the effects of certain impurities on the morphology.

Solutions at rest produced bubbles, around which formed characteristic polycrystalline films (Fig. 2).

I found that the urea concentration could be varied from twice molar to half molar without affecting the growth. The effects varied greatly with calcium salt used. The chloride gave nearly twice as much calcite as did the acetate, for example. It was not possible to grow monocrystals from seeds in solutions at rest. Rotating seeds were used subsequently.

A spherical flask fitted with a sealed axle was used (Fig. 3). The lower end of this axle carried an arm 2, to which two seeds 3 were held by platinum wire. The flask, if thick-walled and well sealed, could be kept at a low pressure. The speed was adjustable from 30 to 60 r.p.m. The first tests showed at once the monocrystals could be grown. Seeds of mean weight 0.25 g grew by $\frac{1}{60} - \frac{1}{30}$ of their initial weights when calcium acetate was used for 10 days. Calcium chloride gave an increase in weight of $\frac{1}{10}$. The acetate gave transparent layers. These layers, if removed from the seeds, left smooth surfaces behind. The external surfaces of these layers were always roughened. The layers given by the chloride were milky white.

These latter tests indicate that calcite monocrystals can be grown, but they have not yet reached the stage where one can envisage growing monocrystals on a larger scale. The seed rhombs in all cases grew quite

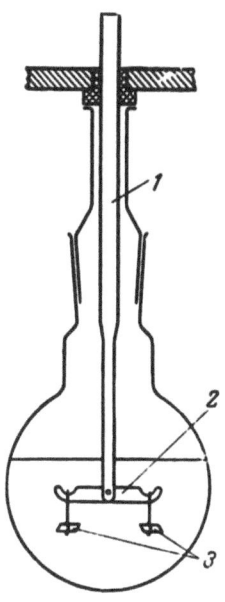

Fig. 3. Rotating-seed crystallizer:
1) axle; 2) arm; 3) seeds.

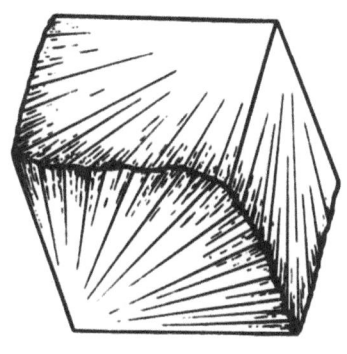

Fig. 4. Shape of rhodochrosite crystal.

regularly. Parasite crystals grew on the vessel, on the arm, and sometimes on the seeds.

Although we learned much from these tests, we found that urease had disadvantages. The main one was that it decomposed quickly. Fresh urease had to be added repeatedly if the tests lasted for more than 10 days. Also, urease is poisoned by heavy metals. It could not be used to grow all the related carbonates for this reason.

I found that urease was activated by certain activators. In some cases these auxoureases were effective even when heavy metals such as copper were present [1].

The urease precipitated at once when salts of zinc, cobalt or cadmium were added to the solution, but the pH did not change. Iron inhibited the urease greatly, and the pH rose only very slowly. It took 10 days to rise from 5.5 initially to 7. Salts of manganese and magnesium had less effect on the urease; the pH rose to 9 in 10 days.

It is clear that zinc, cobalt and cadium carbonates cannot be grown with urease. Magnesium deposited partly as basic and hydrated carbonates at low temperatures. Manganese deposited in small amounts only, as a hydrated carbonate, while iron did not deposite as its carbonate at all.

A hydrothermal synthesis with urea was then tested. Sulfuric acid and caustic soda will decompose urea. These reagents were used because no suitable catalyst could be found. Thick-walled glass tubes were used at 150°C and 5-6 atm. The results began to be apparent after 36 hours.

The sulfates of magnesium, manganese and iron were used to make magnesite, diallogite (rhodochrosite) and siderite. The solutions were mixed with solutions of urea containing a little sulfuric acid. Seeds were not used.

Magnesite crystallized as perfect rhombs, and but little of the material went to form a polycrystalline coating.

Rhodochrosite also crystallized as euhedral rhombs with accessory faces, which gave the crystals the appearance of scalenohedra (Fig. 4). The habit is the usual one for rhombohedral carbonates.

Siderite also formed as rhombohedral crystals. The crystals were ten times smaller than those of the other two minerals, however, although they were highly perfect. X-ray data showed that the crystals were the pure carbonates in all three cases.

Smithsonite was made by using alkali, urea and zinc sulfate. Caustic soda precipitated zinc hydroxide, which then dissolved up as sodium zincate $Na[Zn(OH)_3]$ in excess alkali. Urea was added to the zincate solution, and the mixture was sealed in a glass tube. Little smithsonite was formed, and that as very small spheroids. X-ray tests showed that lines belonging (probably) to zincate were present, as well as smithsonite lines.

Hence magnesite and rhodochrosite can be grown by hydrothermal synthesis, with urea and acid. This has not been done for siderite and smithsonite.

We have here dealt only briefly with this new way of growing carbonates. The work is not yet finished, because other methods have not yet been tried.

I think the work has been successful because it has shown that these carbonates can be grown by chemical crystallization. The rate at which the carbonate is formed is controlled via the rate of decomposition of the urea. Similar methods could be used for the sulfides and sulfates.

Further progress depends on finding suitable catalysts. The method is, however, undoubtedly more convenient than are high-temperature syntheses.

LITERATURE CITED

[1] E. Abderhalden, Handbuch der biologischen Arbeitsmethoden. Abt. IV, Bd. I (Berlin, 1936).

[2] Dana's System of Mineralogy, vol. 2, ch. 1 (Russian translation) (IL, 1953).

[3] C. Doelter, Handbuch der Mineralchemie. Bd. 1 (Dresden-Leipzig, 1912).

[4] C. Hintze, Handbuch der Mineralogie. Abt. I, Bd. 3 (Berlin-Leipzig, 1916-1929).

[5] N. Yu. Ikornikova and V. P. Butuzov, Dokl. Akad. Nauk SSSR 111, 1 (1956).

[6] J. Mellor, Comprehensive Treatise on Inorganic and Theoretical Chemistry.(London, 1946-1948).

[7] G. Rose, Ann. Phys. Chem. 42, 353-398 (1837).

Department of Mineralogy, College of Chemical Technology, Prague

GROWING MONOCRYSTALS OF ANTHRACENE

N. N. Spendiarov and B. S. Aleksandrov

INTRODUCTION

Anthracene is the most efficient organic phosphor at present known; it converts absorbed energy to light quanta with maximum efficiency, and is used as a standard substance abroad. A reliable method of growing anthracene monocrystals is therefore much needed.

The literature on growing anthracene monocrystals is rather scanty. There are five papers that deal directly with this topic.

Ross [1] has grown such crystals by Stockbarger's method in sealed pyrex ampules. The growth rate was 0.8 mm/hr. Crystals up to $3.5 \times 4.5 \times 2.0$ cm were obtained.

Feazel and Smith [2] have used a similar method. The material, sealed in a glass ampule, was lowered in a vertical oven at about 2 mm/hr. The solidified material consisted of three or four crystals about $1 \times 1 \times 2.5$ cm. The growth rate was clearly a little high.

Mette and Pick [3] have detailed methods of purifying anthracene by vacuum distillation, and have made monocrystals as plates 1 mm thick and 3×4 cm in area.

Tarjan [4] has grown anthracene monocrystals 2.3 cm in diameter and up to 5.0 cm long by Bridgman's method.

Finally, Lipsett's group have dealt with growing naphthalene and anthracene [5]. They have described an oven, have indicated the advantages of the seed technique, and have given methods of working the crystals. However, although they say much about naphthalene (for which exceptional results were attained), they say nothing about growing anthracene from melts, and, presumably because the results were bad, they turned to using solutions and vapor to grow this substance.

Very detailed studies of how the surfaces of anthracene crystals age, and of how this effect worsens the light output, [6] show how important these effects are.

In the USSR the Institute of Crystallography, Academy of Sciences of the USSR, and certain other institutes, have been concerned with growing anthracene monocrystals [7].

The Problem

There are some special difficulties in purifying and storing anthracene, and some more several ones in growing monocrystals; these are caused by some properties of the substance.

The C−H bond energies are low, so some hydrogen atoms are readily replaced by radicals. Peroxides are formed photochemically when anthracene is stored in air, and these are an early stage of a reaction that leads ultimately to anthraquinone and other products. The amounts of peroxide and anthraquinone depend on the light intensity and time, and on the grain size of the powder, because the reactions occur in a thin surface layer. It is not sufficient to make the starting material pure enough. The pure material has to be stored until it is used up in growing the crystals, and this storage presents a more difficult problem. Finally, some problems to be solved relate to the actual growing operation (choice of best temperature gradient and shape of crystallization isotherm), and to annealing, extraction and working the crystals.

Purification

We encountered our main difficulties in purifying the material. We always got clusters of crystals separated by dark-colored films of unknown nature when we used coal-tar anthracene of 97% purity, further purified chemically [9], or synthetic anthracene labeled "special pure grade for research purposes." The fact that the crystals ejected this impurity shows that the partition coefficient is much less than one, i. e., that zone purification should be very effective.

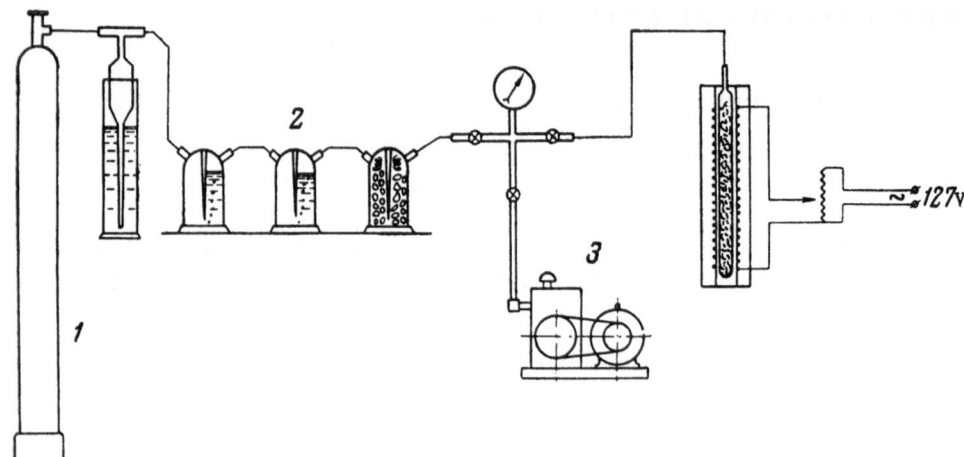

Fig. 1. Line for filling tubes with nitrogen.

In fact it was enough to pass the purified material once through a device with four zone heaters [8] to get material of purity adequate for making good monocrystals.

The zone method is very effective for purifying organic substances. It is enough to treat 97% coal-tar anthracene 16-20 times in this way to get material suitable for monocyrstals.

The most economic method for large-scale use is the following set of operations:

1) the anthracene is distilled with ethylene glycol; the distillate is washed with water containing 10% alcohol, and with 3% soda solution;

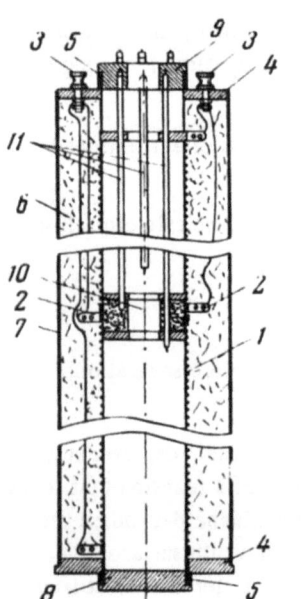

Fig. 2. The oven for growing anthracene monocrystals.

2) the material is then adsorbed on an alumina gel column, with benzene as solvent; heated jackets may be used to speed up this operation;

3) finally, zone purification.

The anthracene is already removed from contact with air at the latter stage. A carefully washed glass tube is loaded with the leafy purified anthracene, and is filled with nitrogen from a line (Fig. 1). This line consists of a cylinder of especially pure nitrogen 1, a system of calcium chloride traps and pyrogallol purifiers 2, and a forevacuum pump 3. A system of taps allows one to pump out the tube while it is heated to 180°C, to fill it to 80-100 mm Hg with nitrogen, and to seal it off.

Growing the Monocrystals

Figure 2 shows the oven used. The quartz tube 1 is wound with nichrome wire 0.6 mm in diameter. The ends of the windings are brought to terminal blocks 2, which are connected to the input terminals 3, which latter are fixed to the top plate 4 of the oven. The cool parts of the oven are protected by the aluminum end-rings 5, which are luted on. The windings are insulated by the asbestos 6, and the hot section as a

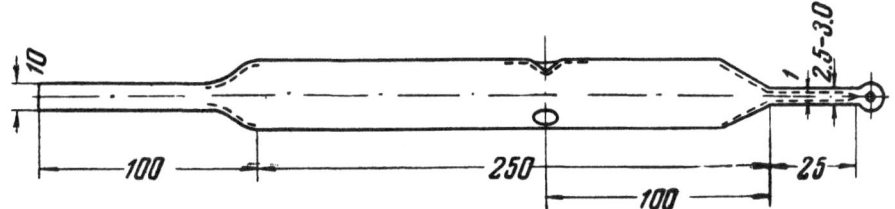

Fig. 3. The tube.

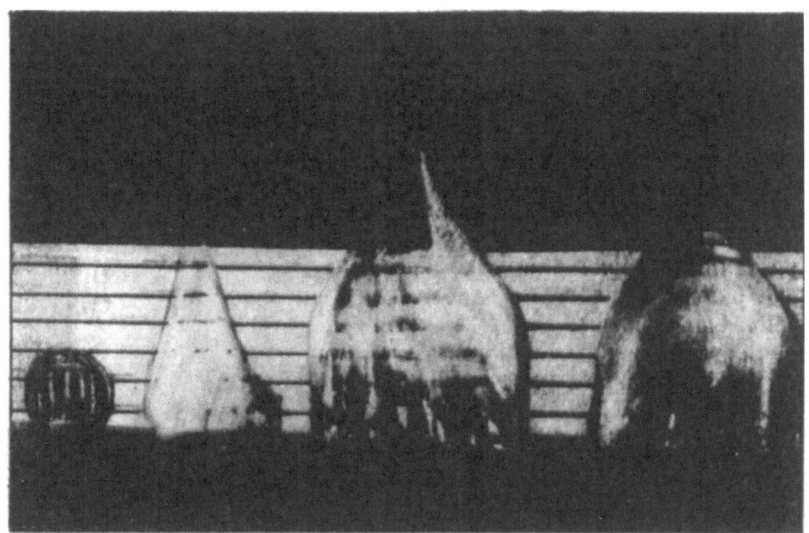

Fig. 4. The first large anthracene monocrystals.

whole is enclosed in the demountable sheet steel case 7, which has two observation holes. The base has an asbestos sheet plug 8, and a top piece 9, to which is fixed the zone heater 10 via two current-carrying rods; this piece also carries the thermocpuole 11. All internal parts can be got at for repair.

A clockwork mechanism is fixed above the oven to the common base; it winds a nichrome wire that supports the tube. The tube moves at 1 mm/hr. The voltage applied to the heaters is stabilized by an SN-2 stabilizer.

The final zone purification is done in the tube in which the monocrystals is to be grown. The purified anthracene then never comes in contact with the air; the tube is not opened, and so can be stored between purification and crystal-growing.

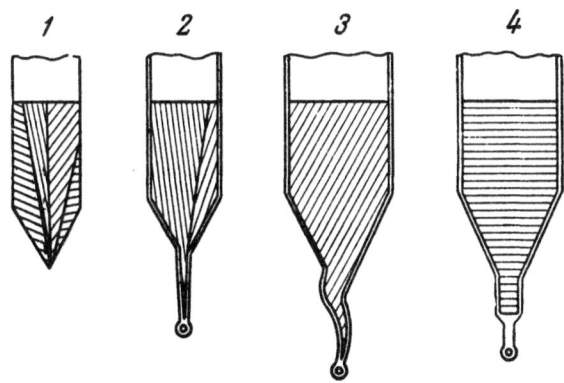

Fig. 5. The most probable growth forms of organic crystals: 1) simple conical end; 2) straight capillary end; 3) spiral capillary end; 4) end with seed.

Figure 3 shows the tube. At one end there is a capillary section, used to make a monocrystal seed, and at the other end there is a pumping tube, which is sealed off when the tube has been filled with nitrogen. The zone purification moves the impurities away from the capillary. Part of the anthracene is melted carefully before the crystals is grown; the impure and is allowed to stay solid.

This impure material must not melt during the growing operation. The top heater is kept at $180 \pm 10°C$, the central one at $235-238°C$, and the bottom one at $180-190°C$ (anthracene melts at $217°C$).

Monocrystals 20 mm in diameter were grown in ovens of this type; so were some early crystals 40 mm in diameter. Figure 4 shows some anthracene monocrystals.

It is essential to see that the material starts growing as a monocrystal. The various shapes of tube shown in Fig. 5 were tested. Straight capillaries gave a 10-15% yield of double crystals. The best shape was one in which the conical part had a vertex angle of 45-60°, and which had a spiral capillary. The start of this spiral (where the nucleus forms) must not be visible from any point within the wide part of the tube; the spiral must make not less than half a turn about the axis of the tube, and must join smoothly onto the conical section.

Freely grown crystals mostly had their cleavage planes parallel to the tube axis (up to 80%). Those with their cleavage planes normal to this axis were more transparent, and were probably better phosphors.

Several crystals were grown from oriented seeds in the tubes shown in Fig. 5, 4. The crystals were highly transparent. The wavy or streaky appearance usual in crystals grown without oriented seeds was quite absent. It is clearly best to use seeds, especially if large crystals are needed.

Annealing

The crystal was loosened from the wall of the tube by a fusion method before it was annealed.

The upturned tube, unbroken, was inserted in a special oven for this purpose; the temperature was above the melting point of anthracene. The result was to leave the crystal lying freely in the tube.

A modified TS-15M thermostatic bath was used for the annealing. The bath contained a holder for the tube; the filling was glycerol, and the heat insulation to the bath had been improved; a control device reduced the temperature at a set rate. The crystal was first kept at 200-210°C for 2-3 hr the temperature was then reduced automatically at 4-5°C per hour.

Working and Mounting

A wire saw was used with anisole (phenyl methyl ether) as solvent to cut up the crystals. The cut surfaces were ground on a lap and were polished with rouge or GOI (State Optical Institute) paste. The crystals were held in an insulating holder to prevent their being heated by the hand.

An anthracene crystal soon becomes coated with yellow oxidation products, which impair the scintillation efficiency. The finished crystals were therefore mounted immediately in sealed mounts. They were held to the glass of the mounts with a silicone fluid of viscosity about $0.5 \cdot 10^6$ centipoise; the mount was coated with magnesium oxide (as reflector), and was then sealed up.

The mounted crystals were tested in a scintillation counter. A single-channel analyzer was used with a Cs^{137} source of activity about $0.05 \mu C/cm^2$.

The efficiency values agreed with published data (e.g., those of [9]), being 140-160% of the efficiency of stilbene crystals of the same size.

SUMMARY

It has been found that purity is essential in growing organic crystals from the melt. Anthracene is especially difficult to use, because it reacts rapidly with air; it has therefore to be kept pure at all stages in the process.

It has been shown that it is very important to orient the growing crystal in order to ensure high quality, if the crystal is one that shows a marked tendency to cleave. The use of oriented seeds is the best way of making organic monocrystals.

The monocrystals produced had scintillation efficiencies that agreed with published values.

No method of mounting the crystals has yet been discovered that will ensure that the efficiency is stable in time.

LITERATURE CITED

[1] Ross, Proc. Nat. Electron Conference 6, 533 (1950).

[2] Feazel and Smith, Rev. Sci. Instr. 19, 817 (1948).

[3] Mette and Pick, Z. Physik 134, 566 (1953).

[4] Tarjan et al., Magyar fiz. folyoiorat 3, 4 (1955).

[5] Lipsett, Canad. J. Phys. 35, 284 (1957).

[6] Lipsett et al., J. Chem. Phys. 26, 1444 (1956).

[7] L. M. Belyaev, B. V. Vitovskii, and G. F. Dobrzhanskii, Growth of Crystals, vol. I [in Russian] (Izd. AN SSSR, 1957).*

[8] N. N. Spendiarov, B. S. Aleksandrov, et al., "Zone purification of anthracene," Trudy IREA (in press).

[9] Nucleonics 6, 5, 70 (1950).

---

* See C. B. translation.

THE ORIENTATIONS OF MONOCRYSTALS OF CERTAIN FERRITES

A. A. Popova

Some goniometric studies are reported for cobalt and manganese ferrites grown by Verneuil's method [1, 2] in an apparatus designed by S. K. Popov [3].

The mixtures from which the ferrites were made consisted of iron-ammonium alum mixed in suitable proportions with the double sulfates of the divalent metals; these mixes were fired in platinum crucibles at 1000°C for 40 min. The sinters were ground up in an agate mortar and were sieved through sieves having 100×100 holes per cm^2.

I grew monocrystals of various sizes in collaboration with A. S. Zavoronkina and I. I. Kargin.

The monocrystals were rounded matt boules or rods of an irregular cylindrical shape; they had slight signs of faces at their tops. The rough rounded look was caused by numerous fine particles adhering to the sides. These minute crystals were so well oriented that they gave a sharp signal; hence accurate goniometer readings were obtained; a similar effect has been found with corundum boules [4]. The crystals used on the goniometer were about $5 \times 4 \times 4$ mm in size. In every case the growth axis coincided with the horizontal axis of the goniometer.

Monocrystals of Co ferrites. The top of a Co ferrite crystal became somewhat overheated when growth ceased, and so tended to run a little. It was therefore difficult or impossible to get signals from the end faces. It was fairly easy to get signals from faces whose zone had its axis coincident (or nearly so) with the growth axis.

TABLE 1

Face no.	$\varphi,°$	$\rho,°$	Signal sharpness (on 10-ball scale)	Symmetry of growth figure
1	0	90	1	—
2	35	90	4	3
3	90	90	7	4
4	145	90	7	—
5	180	90	1	—
6	215	90	3	3
7	270	90	3	4
8	325	90	2	—
9	90	35	2	—
10	270	37	1	—

Table 1 gives the results for one of the three crystals used, and also gives the symmetries of the growth figures seen on certain faces. The other two crystals gave very similar results.

Figure 1 shows a projection of the faces found; it is clear that the crystal belongs to the cubic system and is made up mainly from the simple forms: <111>, <110>, and <001>. The growth axis coincides with one of the axes, L_2. The (111) faces predominate on the cylindrical surface, and give the brightest and sharpest signals; (001) faces are less developed, while (110) faces, though very rare, are always present. The characteristic growth pits seen in Figs. 2 and 3 were seen clearly on the octahedron and cube faces even at low magnifications ($\times 30$,

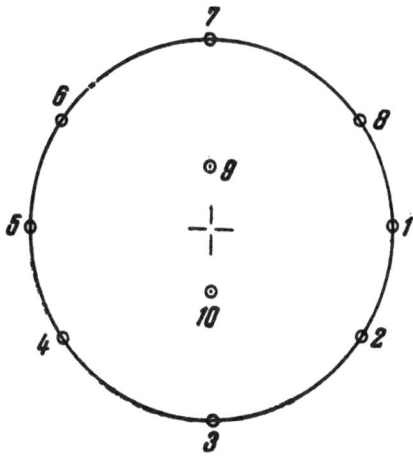

Fig. 1. Projection of a Co-ferrite monocrystal on a plane normal to the growth axis.

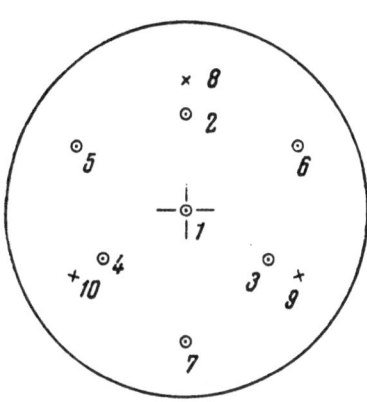

Fig. 4. Growth pits on the (111) face of an Mn-ferrite monocrystal. × 800.

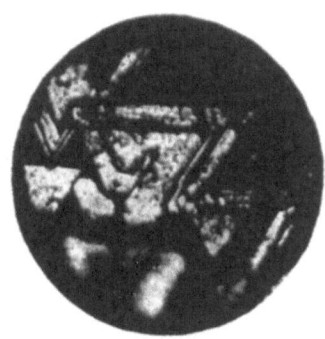

Fig. 5. Projection of an Mn-ferrite monocrystal on a plane normal to the growth axis.

Fig. 2. Growth figures on the (111) face of a Co-ferrite monocrystal. × 30.

Fig. 3. Growth figures on the (001) face of a Co-ferrite monocrystal. × 90.

× 90). Signals were sometimes found from the faces of other simple forms, but these faces are got given in the table, because more careful studies are needed to find their indices.

Monocrystals of Mn ferrites. These crystals had ends of a fairly good shape, without sign of overheating. A careful study in reflected light (× 800) revealed that the faces whose zone had its axis parallel to the growth axis had no growth figures. The faces at the end had good growth figures, however. These figures are large skeletal structures, with a tendency to spiral form (Fig. 4).

Table 2 and Fig. 5 give the results for Mn ferrites. It is clear that two simple forms, <111> and <001>, are present, and that the faces of <111> predominate.

TABLE 2

Face no.	φ,°	ρ	Signal sharpness (on 10-ball scale)	Symmetry of growth figure
1	0	0°	2	3
2	60	55°02′	4	—
3	180	54°42′	9	—
4	300	55°02′	5	—
5	0	70°	5	—
6	120	70°	6	—
7	240	70°	6	—
8	60	109°21′	1	—
9	180	109°30′	1	—
10	300	110°00′	2	—

The growth axis of this crystal coincided with the L_3 axis.

These data show that the orientations of ferrite monocrystals grown by Verneuil's method can be found goniometrically, which should facilitate growing monocrystals of preset orientations.

I wish to thank Professor E. E. Flint for valuable advice on this work.

LITERATURE CITED

[1] A. Verneuil, Compt. rend. 85, 791-794 (1902).

[2] E. I. Scott, J. Chem. Phys. 23, 12, 2459 (1955).

[3] S. K. Popov, Izv. Akad. Nauk SSSR, Ser. Fiz. 10, 5/6, 505-8 (1946).

[4] N. N. Marochkin and I. I. Shafranovskii, Zap. Vses. Min. Obshch., second series, 82, 60-2 (1953).

CRYSTALLIZATION OF FERRITES FROM LIQUID AND VAPOR PHASES

V. A. Timofeeva and A. V. Zalesskii

Several papers have appeared on growing ferrite monocrystals and on the ferromagnetic properties of such crystals [1–4].

The ferrites are, chemically, the salts of the acid $HFeO_2$. The simple ferrites have the formula $M(FeO_2)_2$, where M is any divalent metal. They crystallize in the cubic system and have inversed spinel structures. They are chemically stable, do not dissolve in water or other solvents, and can be made only at high temperatures. At present, ferrite crystals are made as monocrystals from melts, and in polycrystalline forms by cooling solutions of ferrites in fused materials.

The first method is in fact Verneuil's and Bridgman's method and gives monocrystals as fairly large rods and boules (5 × 1 × 1 cm), but high temperatures (1650-1700°C) have to be used.

In the second method two salts are fused together, and one acts as solvent for the other; the temperatures are here much lower (1200-1300°C). The size of any monocrystals so made depends mainly on the solubilities, and on the rate of cooling. The method is finding increasing use for growing monocrystals of many compounds, e. g., of barium and lead titanates and zirconates, of metal oxides, etc. No special pressures or working atmospheres are needed, and crystals of up to 1 × 1 × 1 cm in size can be made [5, 6]. Ferrite monocrystals made in this way have so far not been larger than 2 mm across [1, 6].

The object of the present work was to study the second method, and to examine the ferromagnetic parameters of the crystals. Several fluorides and sodium tetraborate (borax) were tested as solvents. It was found that some fluorides were as good solvents for ferrites as is borax. Manganese and zinc ferrite monocrystals were grown from melts of the corresponding fluorides. The manganese ferrite crystals were 2-3 mm across. The fluorides evaporate rapidly, however. We subsequently used only borax as solvent.

We used No. 10 platinum crucibles in a ShP-1 furnace. The oxides were mixed in appropriate amounts with anhydrous borax. The optimum ratio of borax to oxides was found by trial. The crucible was placed in the cold furnace. The mixture was fused, and was held at a set temperature for a certain time to allow the components to mix; the melt was then cooled slowly.

Monocrystals of some simple ferrites (cobalt and manganese), and of some mixed ferrites (zinc-manganese and zinc-nickel), were made as flat plates (Fig. 1). The method was unsatisfactory, however, because the crystals grew at random, especially at the bottom of the crucible, and so were very difficult to extract.

Salts dissolve in one another in the same way as they dissolve in water. We may thus expect that the ways in which crystals grow from aqueous solutions will also occur with fused salts. Hence there are two ways of growing ferrites from seeds inserted in borax melts: 1) by

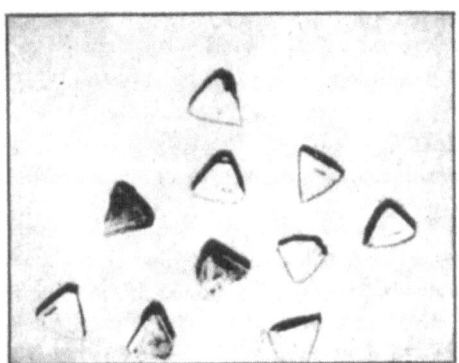

Fig. 1. Monocrystals of manganese and cobalt ferrites grown from melts. × 3.

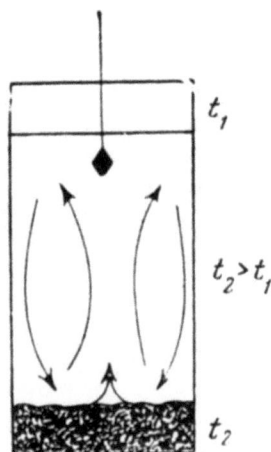

Fig. 2. Vessel for growing monocrystals.

Fig. 3. Crystals of cobalt ferrite grown from a melt. × 3.

Fig. 4. Crystals of cobalt ferrite grown from the vapor. × 2.

cooling saturated solutions, and 2) by evaporating off the solvent at constant temperature. We found that both methods could be made to give ferrite monocrystals.

A special vessel (Fig. 2) was used for the first of these methods. A temperature gradient was produced between the upper and lower parts of the vessel, in order to favor growth on a seed placed in the upper part. In this way we made cobalt ferrite monocrystals as a cluster of crystals, in which cluster the first layer consisted of small crystals, among which certain individuals began to predominate as growth went on; several crystals were obtained as octahedra with sides 6-7 mm long. In some cases only one large crystal finally resulted (Fig. 3). The large crystals were less perfect than the small ones, because the large grew by skeletal-spiral mechanisms in the temperature gradient, whereas the small ones grew under more nearly equilibrium conditions, and were regular octahedra.

We studied the solvent evaporation method, and found that the high vapor pressures of the ferrites made it possible to grow monocrystals from the vapor. The ferrite evaporated as well as the solvent, and deposited as a boule on the edges and walls of the crystallizer. It is known that some spinels can grow from the vapor when fluoride is present [7]. Here the substance that aids the process is elementary boron, formed by decomposition from the borax.

Monocrystals of cobalt and manganese ferrites were grown from the vapor state. The crystals were sometimes 15-20 mm long (Fig. 4).

The x-ray patterns showed that the rods so grown were monocrystals, in whole or in part; the specific gravity (5.2) was close to the value for polycrystalline cobalt ferrite. The crystals had a hardness of 6-7 (Mohs) and could be worked easily.

The crystals were tested magnetically, in order to check their compositions, and to detect any changes in structure caused by heating. The way the permeability varied with temperature was recorded for several crystals. These data gave the Curie points. The values were compared with published data [8] for polycrystalline ferrites. The Curie point is very sensitive to changes in composition, so it can be used to check the composition. Magnetization isotherms were taken to find the saturation induction.

The magnetization was measured by ponderomotive methods. A specimen weighing a few milligrams was suspended on a quartz fiber in an oven with a bifilar winding, and was set in a field of strength H and of defined gradient kH/dx, which had been measured. A VP-NV-20 balance was used to measure the force acting on the specimen. The specific magnetization (per gram) is given by

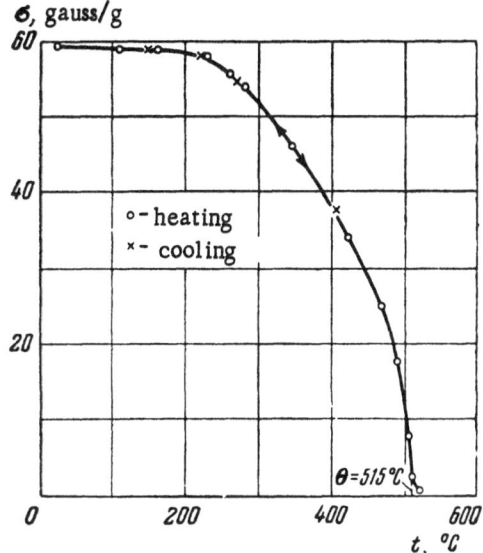

Fig. 5. Temperature variation in the specific magnetization σ for a CoO·Fe$_2$O$_3$ crystal grown from the melt. θ is the Curie point; strength of external field 1340 oersteds.

$$\sigma = \frac{F}{m\,dH\,/\,dx},$$

where F is the force acting, m is the weight, and dH/dx is the field gradient.

Figures 5 and 6 show how the specific magnetization varies with temperature for cobalt, manganese, and zinc-manganese ferrites.

Figure 7 shows data for two cobalt ferrite crystals made in different ways (from a melt and from the vapor state). The values are very similar (the Curie points differ by 15°C only). The differences may be caused by the crystals grown from the vapor being not strictly stoichiometric in composition.

The curves were all very reproducible in repeated heating and cooling cycles, which fact shows that no structure or composition changes occurred. The σ(T) curves have shapes that show that no ferromagnetic impurities are present (such impurities are often found in ferrites).

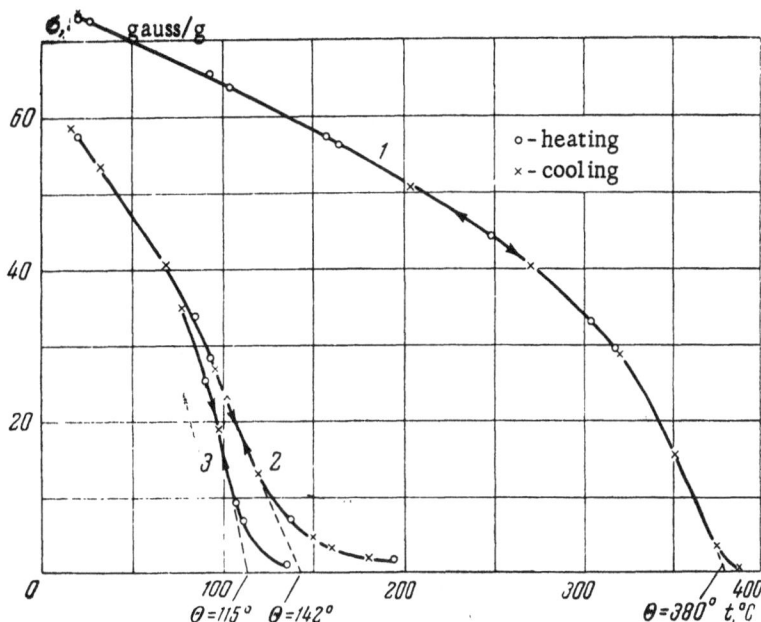

Fig. 6. Specific magnetization σ vs. temperature for ferrite crystals grown from melts: 1) MnO·Fe$_2$O$_3$, 2) 1.5 ZnO·3.5 MnO·5 Fe$_2$O$_3$; 3) 2.2 ZnO·2.8 MnO·5 Fe$_2$O$_3$. Strength of external field 1340 oersteds; θ is the Curie point.

Figure 8 shows magnetization isotherms for a cobalt ferrite crystal (that of curve 1, Fig. 7). Saturation was not reached at room temperature, but a rough estimate (80-100 gauss/g) agrees with values found for polycrystalline specimens.

The table gives our Curie-point values, θ$_m$, for monocrystals, and published [8] values for polycrystalline specimens, θ$_p$.

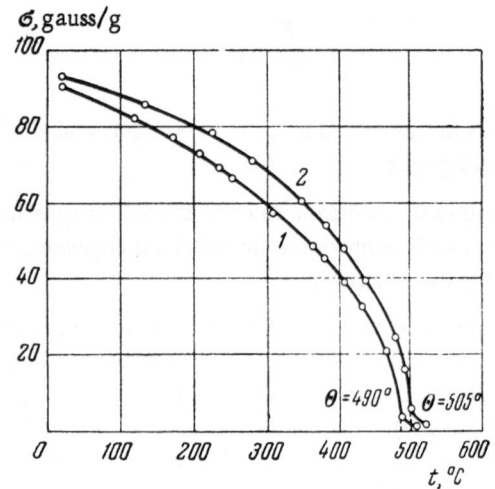

Fig. 7. Specific magnetization σ vs. temperature for CoO · Fe$_2$O$_3$ crystals: 1) grown from the vapor; 2) grown from the melt. External field 4760 oersteds; θ is the Curie point.

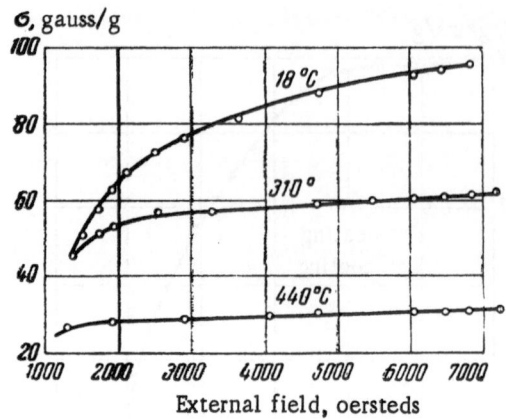

Fig. 8. Specific magnetization isotherm for CoO · Fe$_2$O$_3$ grown from the melt.

The cobalt ferrite gives the best agreement. The Curie points for the other ferrites disagree, perhaps because the crystals had compositions differing from those

Composition	θ_p, °C	θ_m, °C
MnO · Fe$_2$O$_3$	300	380
1,5 ZnO · 3,5 MnO · 5Fe$_2$O$_3$	∼200	142
2,2 ZnO · 2,8 MnO · 5 Fe$_2$O$_3$	∼145	115
CoO · Fe$_2$O$_3$ (from melt)	515	510
The same	515	505
CoO · Fe$_2$O$_3$ (from vapor)	515	490

appropriate to the original mixtures. Qualitative analyses showed that these ferrites contained a certain amount of Fe^{2+} and Mn^{3+} ions, which would shift the Curie points. The properties and compositions of polycrystalline ferrites (especially mixed ones) are very sensitive to the method of manufacture (sintering atmosphere, heat-treatment).

Timofeeva dealt with growing the ferrite monocrystals, and Zalesskii examined the magnetic parameters.

The authors wish to thank Professor K. P. Belov for scientific guidance, and engineer L. D. Prokhorov for assistance in the experiments.

LITERATURE CITED

[1] J. Galt, B. Matthias, and J. Remeika, Phys. Rev. 79, 391 (1950).

[2] T. Okamura and J. Kojima, Sci. Rep. Research Inst. Tohoku Univ. A4, 72 (1952).

[3] J. Smiltens, J. Chem. Phys. 20, 6, 990 (1952).

[4] E. Scott, J. Chem. Phys. 23, 12, 2459 (1955).

[5] B. Matthias, Phys. Rev. 75, 11, 1771 (1949).

[6] J. Remeika, J. Chem. Soc. Amer. 76, 5, 940-941 (1954).

[7] V. A. Timofeeva and I. I. Yamzin, Trudy Inst. Krist. No. 12 (1956).

[8] C. Guillaud, J. Phys. radium, 12, 3, 239-248 (1951).

THE GROWTH OF BARIUM TITANATE CRYSTALS FROM A
BARIUM CHLORIDE MELT

V. A. Timofeeva

The $BaCl_2$–$BaTiO_3$ system has been examined [1], and it has been found that barium titanate monocrystals can be grown within broad temperature and concentration limits.

We deal here with the growth processes of $BaTiO_3$ crystals under various conditions of heating and cooling, and with varying compositions in the original mixtures. The methods have been described already [1].

We began by studying how the system behaved as it was heated to the melting point.

The charge was made from barium carbonate, titanium dioxide and barium chloride; the two former components gradually sank to the bottom of the crucible, because they are the denser ones. The barium titanate that formed as the temperature was raised then diffused into the fused barium chloride.

We found that supercoolings of 80-100°C were needed to cause the $BaCl_2$–$BaTiO_3$ system to deposit barium titanate if no seed was present. Hence seeds are needed to grow $BaTiO_3$ crystals.

There are, however, technical difficulties in inserting a seed into a melt of highly volatile barium chloride; these have not yet been overcome.

The best and easiest method is to let the melt cool without stirring it, when it crystallizes from the bottom upwards, i. e., from the most supersaturated area. If the distribution in height of $BaTiO_3$ in $BaCl_2$ can be arranged to be the same as that for the crystallization curve, it should be possible to grow the crystals right down to the eutectic point. Then the size of the crystals will be determined by the depth of the layer of melt.

Figure 1 shows this scheme graphically. On the left we have a vertical section of the crucible, on the right the melting-point curve for the $BaCl_2$–$BaTiO_3$ system. The top layer consists of pure $BaCl_2$, and the bottom one is almost all $BaTiO_3$. A regular gradation in barium titanate concentration with height is a basic essential if the $BaTiO_3$ crystals are to grow uninterruptedly.

Figure 2 shows one such melt, where the barium chloride is white, and the titanate (wing structures) is dark. The $BaTiO_3$ crystals that began to grow at the bottom have reached the top, and are of a size equal to the depth of the melt. Geometrical selection has caused only those crystals for which conditions were favorable to grow to large sizes.

The crystals take the curious wing shape seen in Fig. 2 because they twin when they form from fused salts [2]. Above 1450°C barium titanate exists in a hexagonal modification (which is not ferroelectric). Below this temperature it is cubic. We have found that this titanate crystallizes from barium chloride in the hexagonal form under certain conditions. The crystals are thin transparent hexagonal plates if the $BaTiO_3$ predominates greatly over the $BaCl_2$ (near the $BaTiO_3$ ordinate in Fig. 1).

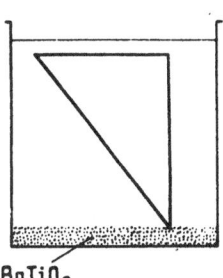
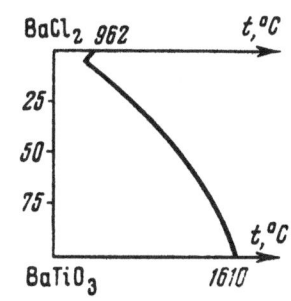

Fig. 1. Scheme for growing barium titanate monocrystals in accordance with the crystallization curve for $BaTiO_3$–$BaCl_2$.

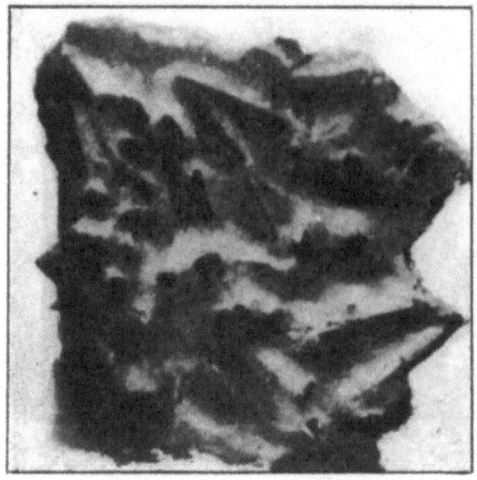

a

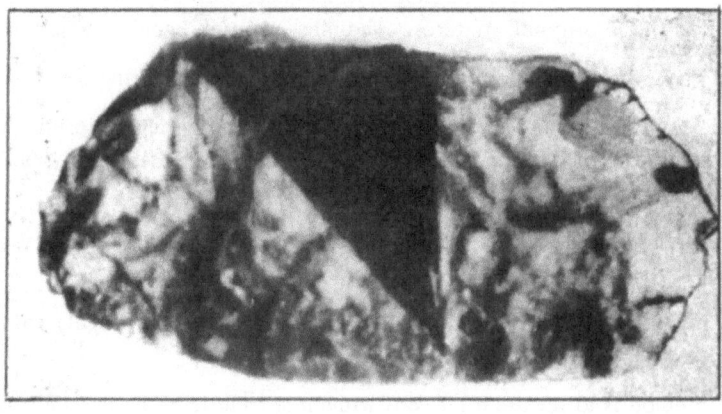

b

Fig. 2. BaTiO₃ crystals grown from BaCl₂. × 2: a) view from above, b) view from the side.

Fig. 3. BaTiO₃ dendrites.

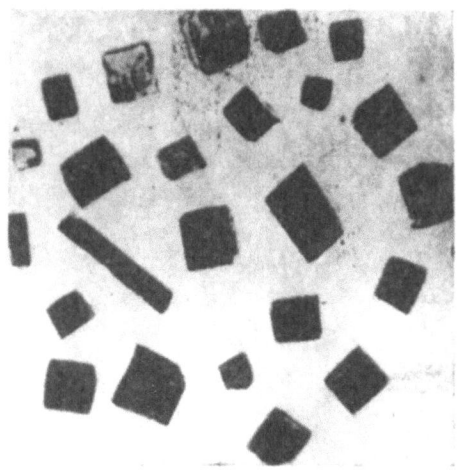

Fig. 4. BaTiO₃ crystals in cubes and parallelepipeds. × 2.

Now the crystals start to grow from the bottom, i. e., from the point where the supersaturation is greatest, and so the hexagonal form is favored. It may be that the transition to the cubic form is the cause of the twinning. The BaTiO₃ twins are polysynthetic. The twin-planes are (100) cube faces.

A notable feature of these twins is that they grow quickly from supersaturated solutions. This shows why it is possible to get small (1-2 mm) thin triangular crystals even when the melt is allowed to cool freely to room temperature from 1400°C. We have found that the plates grow more slowly as the cooling rate is reduced; ultimately only isometric BaTiO₃ crystals are formed. In one case two identical mixtures were heated identically, but were cooled at two different rates, one designed to favor twins (rate 20°C per hr) and the other smaller than the first by a factor of 4-5. In the first case we found

Fig. 5. Spiral growth on a (100) cube face. × 30.

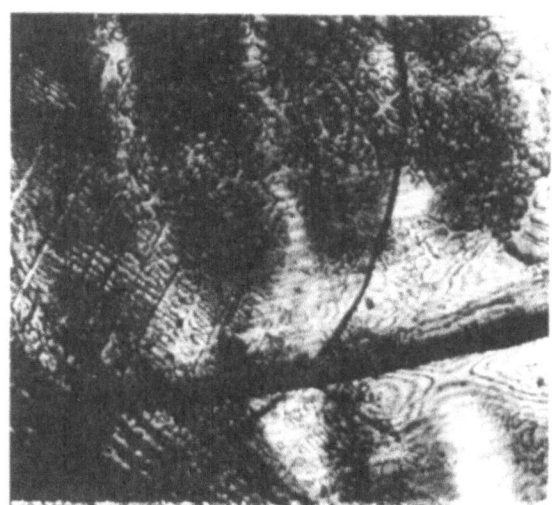

Fig. 6. Surface of a BaTiO₃ crystal grown at a large supersaturation. × 30.

twins, as we expected, and in the second only isometric crystals. The thickness of the wings in the twins varied from a few tenths of a millimeter to over 1 mm, as a function of cooling rate. Dendrites formed (Fig. 3) if the supercooling (supersaturation) was large. If the supersaturation was small (near-equilibrium conditions), the crystals were isometric, being cubes or rectangular parallelepipeds (Fig. 4). Cube faces were usually accompanied by octahedron and rhombohedron faces; in one case the crystals showed tetrahexahedron faces. The crystals can be mace of any shape by controlling the cooling (growth rate), e. g., as thin or thick plates, as regular cubes, or as other regular forms in the cubic system.

The surfaces of crystals grown under near-equilibrium conditions showed clear growth spirals (Fig. 5). In other cases the faces were uneven and covered with dendrites (Fig. 6). More detailed results will be given in a forthcoming publication.

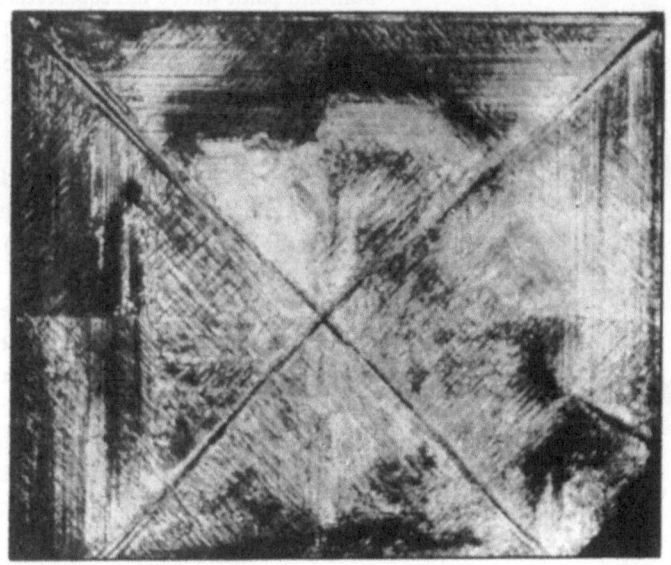

Fig. 7. Growth pyramids and domain structure of a BaTiO₃ crystal. × 30.

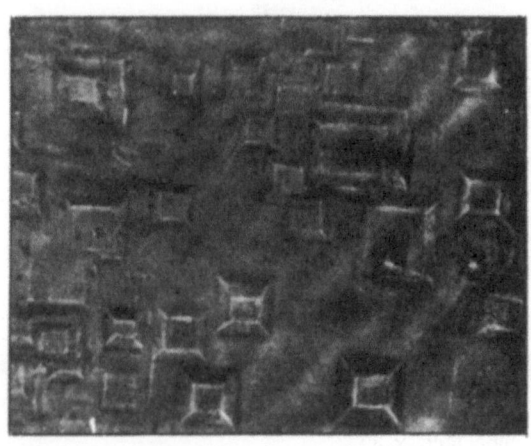

Fig. 8. Etch figure on a (100) cube face. × 80.

Figure 7 shows a domain structure. The surface of BaTiO₃ may be etched with fused BaCl₂ (secondary heating) to give etch figures that are regularly placed cubes. At high magnifications one can see tetrahexahedron faces as well (Fig. 8).

To sum up, we may say that the heating and cooling rates have large effects on the growth of BaTiO₃ crystals from solution in fused BaCl₂. Detailed studies of the conditions under which the various crystal forms of BaTiO₃ appear can alone enable us to control the growth.

I wish to thank by supervisor, I. S. Zheludev, and my colleagues L. D. Prokhorov, R. A. Voskanyan, A. I. Malyshev, I. A. Pleteneva, and N. A. Anisimov, for their advice and assistance.

LITERATURE CITED

[1] V. A. Timofeeva and I. A. Pleteneva, Kristallografiya 3, 2, (1958).

[2] E. White, Acta Cryst. 8, 12, 845 (1955).

A CRYSTALLIZER FOR GROWING ORGANIC CRYSTALS FROM THE MELT

L. M. Belyaev, G. S. Belikov, and G. F. Dobrzhanskii

We [1] have proposed a modification of Stöber's method [2] for growing tolane and naphthalene crystals. The crystallizer has a flat cooled bottom and is placed in a thermostatted bath of liquid.

The crystal grows from an oriented seed that covers the bottom completely. The substance cannot distill off because the crystallizer is closed by a heated lid whose temperature lies somewhat above the melting point of the crystal. This device has the disadvantage that the lid must be lifted to see the crystal, and that this disturbs the crystallization. The lid is difficult to make if a viewing port is included.

The plant described below is meant for growing monocrystals from organic compounds that melt below 100°C. The crystal can be seen without disturbance to the conditions. Figure 1 shows the system. There are two glass cylindrical vessels, one within the other. The outer is filled with a transparent liquid (water, glycerol, medicinal paraffin) and acts as the thermostat bath. The inner is the crystallizer proper. The outer is placed in a tin-plate holder on an electrically heated plate. A contact thermometer and a relay control the temperature. The outer vessel is closed by a plastic-bonded lid with holes in it; these holes serve to admit the crystallizer, the contact thermometer, a direct-reading thermometer, and the coolant tubes. The cooler (a hollow copper vessel with copper tubes placed as far as possible from the crystallizer) is placed within the bath. The cooler has the same diameter as the crystallizer. The crystallizer is clamped tightly to the cooler by a special ring. The crystallizer is closed by a special lid, which is heated to the melting point of the compound during crystallizations. Running water passes through the cooler. A rotameter controls the flow. The cooler removes enough heat to allow one to grow crystals 50-60 mm high. We have used plant jars 22 cm high and 17 cm in diameter, or various aquarium tanks, for the outer vessels. The inner vessels were various battery jars.

The process of crystallization is as follows.

The temperature is first brought to a value 4-5°C above the melting point of the substance. The inner jar is then filled $1/3$ to $1/2$ full with melt. A prepared seed

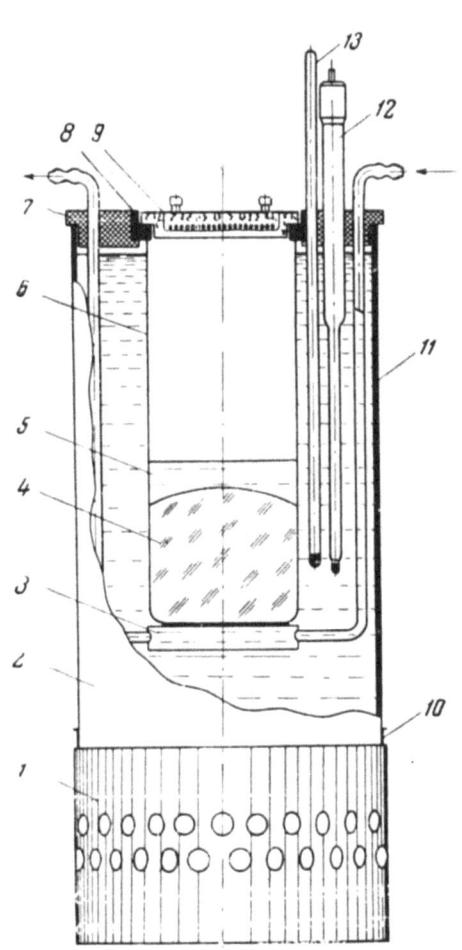

Fig. 1. The crystallizer system: 1) heated plate, 2) contact liquid, 3) cooler, 4) crystal, 5) melt, 6) crystallizer, 7) lid, 8) ring, 9) heated lid, 10) support, 11) thermostat, 12) contact thermometer, 13) direct-reading thermometer.

4-5 mm thick is mounted on a thin copper plate, and is heated very slowly almost to its melting point; it is then lowered into the crystallizer. The seed is prevented from melting away by passing water through the cooler. The growth rate is controlled in terms of the water flow, and by reducing the temperature slowly. The crystal is viewed through the liquid in the thermostat. The cooling is gradually brought to a finish when the crystal has reached the desired size, and the temperature is then raised by 2-3°C. The crystal melts at the sides and can be lifted out on the plate that held the seed.

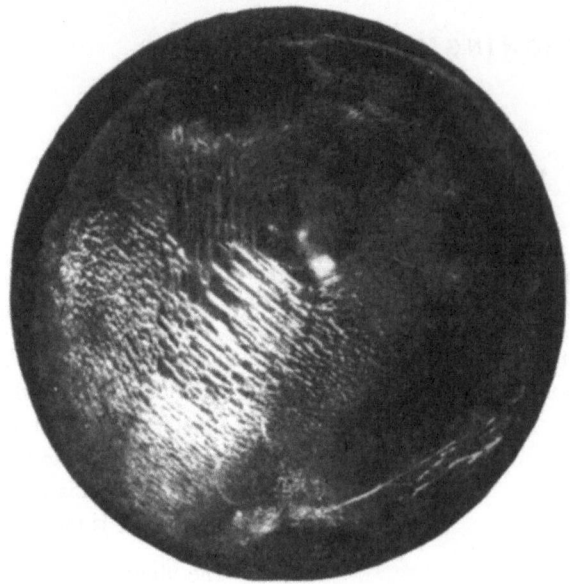

Fig. 2. The surface of a tolane crystal.

We have been able to grow crystals with good faces at the top, which shows that the crystallization has been done properly. Figure 2 shows a tolane crystal; the faces are clearly seen. We have grown monocrystals of di- benzyl (m. p. 50°C), of tolane (m. p. 61°C), and of naphthalene with various additives (m. p. 80°C) in this device. Similar crystallizers have been used at the Kharkov Chemical Reagents Factory for growing tolane and naphthalene crystals.

The system may be used to grow crystals by the Kyropoulos method [3]. Here the crystallizer is held in a base made of a poor conductor of heat. The cooler is a brass or copper rod inserted through the lid. The advan- tage of our system is that the crystal can be seen clearly.

LITERATURE CITED

[1] L. M. Belyaev, B. V. Vitovskii, and G. F. Dobrzhanskii, Growth of Crystals. I [in Russian] (Izd. AN SSSR, 1957).

[2] F. Stöber, Z. Krist. 61, 299-314 (1924).

[3] Kyropulos, Z. anorg. u. allgem. Chem. 154, 303 (1925).

A NEW TYPE OF PISTONLESS COMPRESSOR FOR PRODUCING VERY HIGH GAS PRESSURES

S. S. Boksha

High pressures, produced in solids, liquids or gases, are now much used in physics and chemistry. Artificial diamonds [1] and borazon [2] (a cubic form of boron nitride, nearly as hard as diamond) are made at high pressures. Many chemical processes, e. g., the syntheses of methanol and ammonia, the polymerization of ethylene, etc. are also carried out at high pressures.

The design and construction of special equipment is one of the main difficulties in work at high pressures. High gas pressures are the most difficult to produce, although gas reactions are of the greatest scientific and technical interest.

Several ways of producing high gas pressures are known. Bulk static compression (see [3], pp. 220-233) is the method most often used. Piston, rotary and membrane pumps, and pressure amplifiers (silent compressors) use this method. The latter two types of pump are useful only at fairly low pressures. Piston pumps and amplifiers are alone used at high pressures, but they have serious disadvantages (the gas is contaminated by lubricants, the outputs are small, and very high pressures are unattainable, because moving seals are used *).

Another method which is much used is to compress the gas dynamically in two stages. The gas is set moving at a certain speed in the first stage; the kinetic energy is transformed to pressure energy in the second stage. Centrifugal, axial-flow and injection compressors use this principle (see [4], pp. 117-165). These compressors are used when large outputs are needed at fairly low pressures.

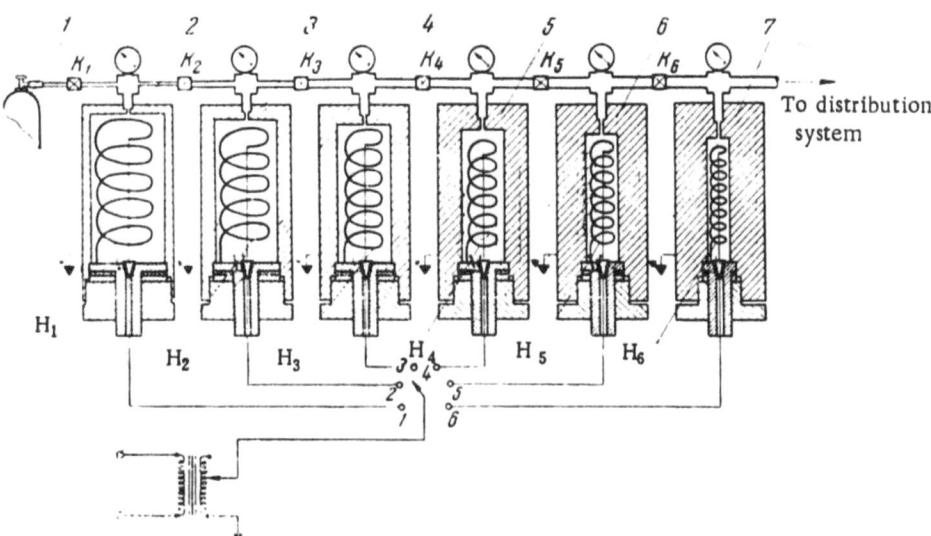

To distribution system

Fig. 1. Schematic diagram of the pistonless gas compressor.

---

* No known industrial compressor works above 1000-2000 atmos, and laboratory compressors (e. g., Vereshchagin and Ivanov's [12]) do not go above 5000-6000 atmos.

Changes of state are much used to produce high gas pressures. Thermal gas compressors (see [4], pp. 166-7) work on this principle. An apparatus for liquefying the gas is needed, as well as the evaporator. If only small outputs are needed, as in some laboratory equipment, the two units may be combined (see [4], pp. 168-171, and [5]). Such devices are called "thermocompressors." They have the advantage that the gas is not contaminated by lubricants.

The above devices have disadvantages. They all demand expensive coolants (liquid nitrogen, hydrogen, helium, etc.) in large amounts, which makes them expensive to work. They have very small outputs, because of the thermal inertia in the large steel cylinders. They are therefore not used at all industrially, and seldom in laboratories, as the few publications on gas compressors indicate [4-7].

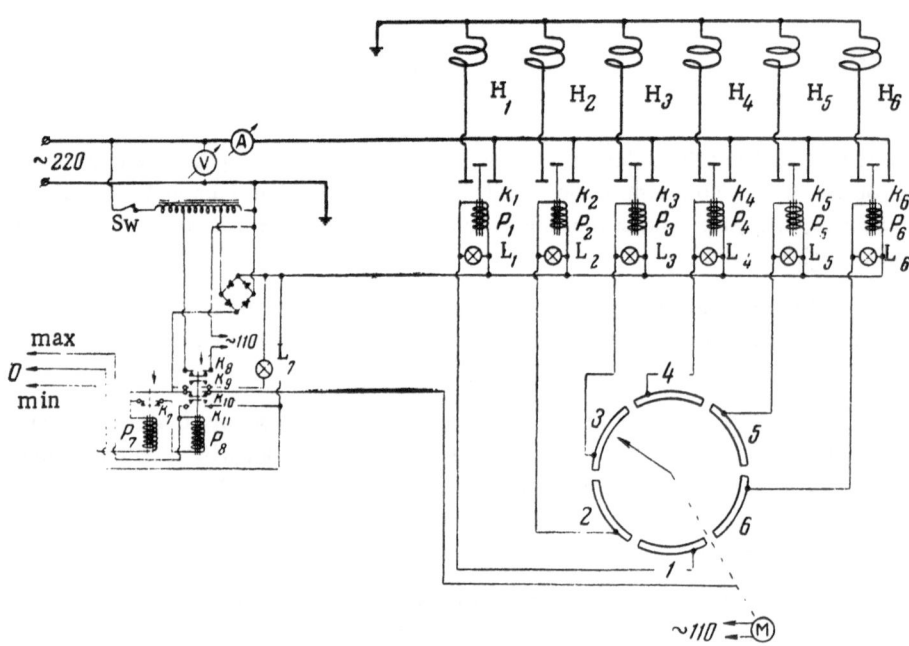

Fig. 2. The electrical system of the compressor.

Our new method [8] is free from many of the above disadvantages. We have designed a pistonless compressor that works up to 3000 atmos which uses this method. Figure 1 shows the principle of the compressor, where the numbers 2-7 denote the high-pressure cylinders, whose volumes are, respectively: $V_1 = 3000$ cm^3, $V_2 = 2000$ cm^3, $V_3 = 1000$ cm^3, $V_4 = 500$ cm^3, $V_5 = 500$ cm^3, $V_6 = 300$ cm^3. These cylinders are joined by steel capillaries with valves (K_1-K_6) that allow the gas to flow from left to right only. Each cylinder is fitted with an internal heater spiral. The entire unit is set in a tank of running water. The valve on the gas source 1 is opened, and the gas fills all the cylinders to a pressure roughly equal to that of the source. The heater H_1 is then switched on in cylinder 2. The gas expands, and in part passes over to cylinders 3-7; the pressure rises.* Then H_1 is switched off, and H_2 is switched on. The gas in 3 is heated, expands, and in part enters cylinders 4-7, raising the pressure further. The process is repeated with the other cylinders. The first cycle ends when the gas in 7 has been heated. By this time the gas in 2 has cooled completely, and its pressure has fallen below that of the source 1. Some gas therefore enters 2, and the second cycle starts; the gas in 2 is heated, expands in part into 3, and so on, until the gas in the last cylinder has again been heated, when the third cycle starts, etc.**

The compression cycles are fully automatic, and are controlled with the circuit shown in Fig. 2. Here, as in Fig. 1, H_1-H_6 denote the heaters. The relays P_1 to P_6 switch the heaters. The signal lamps L_1 to L_6 are wired in parallel with the heaters. The relays are controlled by the switching unit via contacts 1-6, which are driven by the synchronous electric motor M.

* The hot gas cools down almost to the temperature of the flowing water as it passes through the capillaries, because its thermal capacity is small relative to that of the steel capillary.
** At no point is the gas at a temperature below that of the water. The maximum temperature in the cylinders is determined by the heater. Here the temperature does not on the average exceed 400°C.

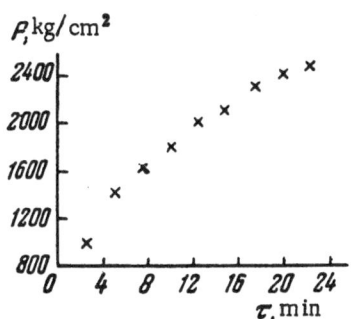

Fig. 3. Presssure-time curve at the output of the compressor.

Relays P_7 and P_8 serve to keep the pressure in the distribution system at a preset value, which is ensured by a system (not shown) with a gauge and adjustable contacts marked max -O- min.

Contact O is fixed to the needle of the gauge. Contact O touches the max contact when the pressure rises to the set limit; the compressor heater is switched off, and lamp L_7 lights. Conversely, the min contact is touched when the pressure falls to the lower limit; the compressor starts up, and L_7 goes out. The heaters are of low inductance, and so the starting torque is small, which is a distinct advantage over piston compressors.

The compressor has been used repeatedly. The working gases have been nitrogen and argon. Figure 3 shows how the pressure rose in a distribution system of volume 100 cm^3 during a test. It is clear that the pressure rose very rapidly during the first cycle. The rise was almost 40% of the total rise. The rise becomes less with each successive cycle, and approaches some limit determined by the volumes and number of the cylinders, by the temperatures used, and by the source pressure. The pressure rose from 150 to 2600 atmos (by nearly a factor 17) in 25 min. The total power consumed was then about 7kw.

Our mechanic É. P. Zukovskii assisted in this work.

LITERATURE CITED

[1] ASEA's tidn. 47, 6-7, 93-94 (1955).

[2] New Scientist 15, 16-18 (1957).

[3] M. P. Bukalovich and I. I. Novikov, Technical Thermodynamics [in Russian] (Gosenergoizdat, 1955).

[4] B. A. Korndorf, High-Pressure Techniques in Chemistry [in Russian] (Gostekhizdat, 1952) pp. 117-165.

[5] Simon, Ruheman, and Edwards, Z. Phys. Chem. 6, 331 (1930).

[6] Ya. S. Kan, J. Tech. Phys. (USSR) 13, 1156 (1948).

[7] J. Robin and B. Vodar, J. phys. radium 17, 500-501 (1956).

[8] S. S. Boksha, Kristallografiya 2, 1, 198-200 (1957).

[9] L. F. Vereshchagin and V. E. Ivanov, Pribory i Tekhnika Eksperimenta No. 4, 73-7 (1957).

THE GROWTH AND MORPHOLOGIC SYMMETRY OF
BENZOPHENONE CRYSTALS

A. A. Chumakov

We have worked on laboratory methods of growing fairly large uniform monocrystals of benzophenone in order to study some physical properties of this compound.

There are several papers on growing benzophenone monocrystals from the melt. An apparatus [1] and a method of using supercooled melts have been described. It is stated that a transparent monocrystal of volume 2 cm³ has been grown in this way.

Others [2] have made crystals 2-3 cm long along the c axis in face growth-rate studies. It is also stated that benzophenone monocrystals can be grown from solutions in alcohol [3], ligroin [4], benzene and acetone. In all these cases large and perfect monocrystals were not the main objective, however.

We have found that solutions have advantages with benzophenone, and so we have given our attention to methods of using solutions.

Benzophenone dissolves in many organic solvents. Good solvents are ether, chloroform, benzene, acetone, dichloroethane, carbon tetrachloride, benzine, alcohol, etc. We have tested all the above solvents, and have found that benzine, acetone, and carbon tetrachloride are the best. Figure 1 gives solubility curves for the latter two solvents.

The purest and most perfect crystals are to be had only from thoroughly purified materials. It is very difficult to purify benzophenone [5]; we had to recrystallize the material many times from solution in order to get the desired purity. Purity in the solvent is also essential, because it affects the solubility and the quality of the crystals. We therefore used the purest solvents.

We used water baths with Hoepler-type ultrathermostats, modified in accordance with some suggestions of V. F. Parvov, to grow the monocrystals. Glass vessels were used instead of metal ones, so the crystals could be seen at all times. The temperature was kept constant to ± 0.01°C within the limits +20 to +40°C. Special attention was paid to sealing the crystallizer. This sealing is especially important with volatile and flammable solvents such as acetone and benzine.

The crystal-holder was turned continuously in one direction, or was oscillated. The speed was varied from 200 to 400 rpm. Special baffles were fitted to prevent the solution from swirling and to trap dust and minute crystals that might appear in the solution.

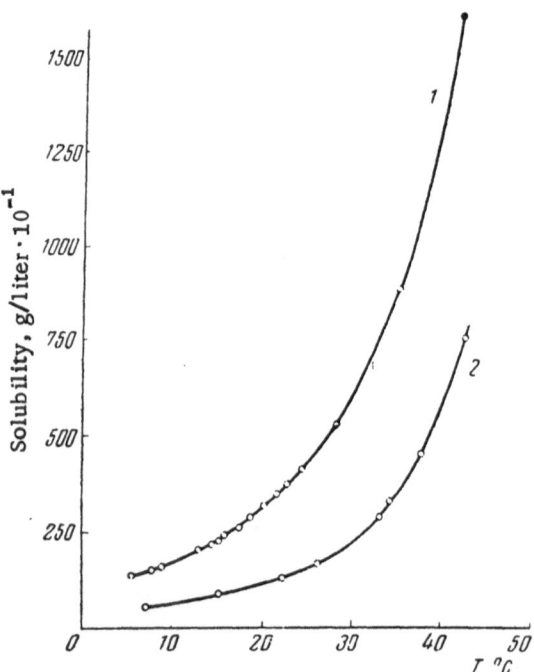

Fig. 1. Solubility curves for acetone (1) and carbon tetrachloride (2).

Fig. 2. Thermostat and crystallizer.

Fig. 3. Benzophenone crystal grown from acetone.

Figure 2 gives a general view of the apparatus.

Monocrystals of volume up to 300 cm^3 could be grown in a 3000 cm^3 crystallizer (Fig. 3), because benzophenone is so soluble in acetone and carbon tetrachloride. The total drop in temperature during one crystallization cycle was only 1°C. The solution was cooled to 0.1°C below the saturation point. Large supersaturations

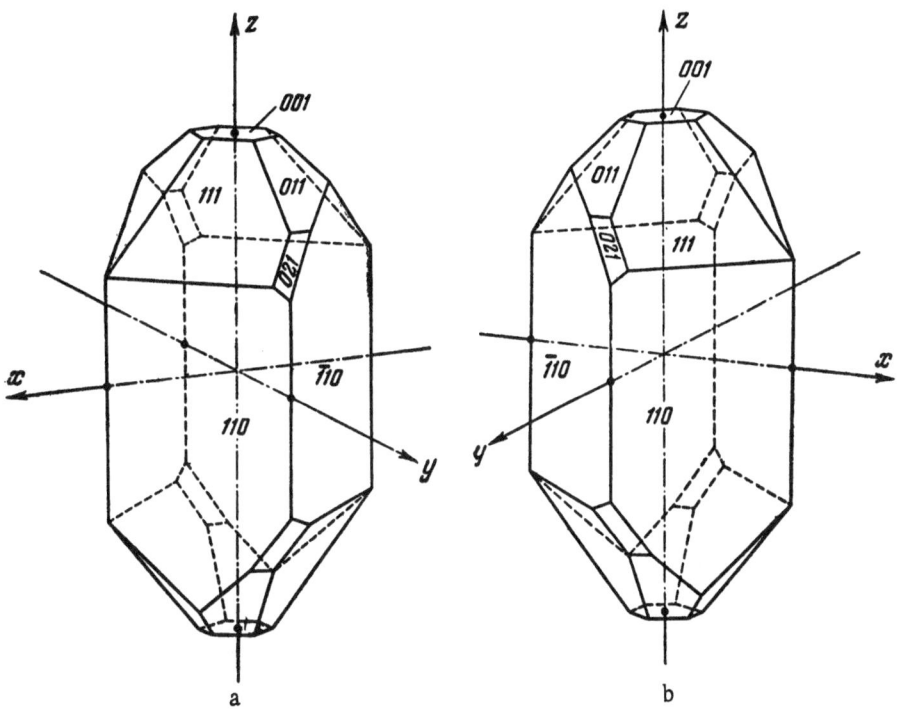

Fig. 4. Ideal growth forms of (a) right-handed and (b) left-handed benzophenone crystals.

caused all the faces, except {110} (rhombic prism) to be cloudy, even at speeds of 400 rpm. The best temperature range for use with solutions in benzine was 20-25°C. A wider range (20-40°C) was usable with acetone and carbon tetrachloride. The conditions approached those in the melt, because benzophenone is so soluble in acetone and carbon tetrachloride. Above 40°C the solvents are too volatile, especially in the early stages of crystallization.

All surfaces defects (cracks, cloudy areas, etc.) heal up rapidly and completely if the conditions are correct; the crystal grows transparent, with perfect faces. The habit depends on the solvent [6]. The effect is clearly seen with benzine and acetone. Crystals grown from benzine are very much elongated along c, and have to be cut back repeatedly if large ones are needed. Acetone and carbon tetrachloride give much higher relative growth rates on the {110} prism faces, and so the crystals approach much more nearly the ideal shape shown in Fig. 4.

Now, the {110} prism faces grow more slowly than the others, and os it is best to rotate the crystal in the solution about the c axis. The prism faces then encounter the best growth conditions, and the crystal becomes more nearly isometric.

We found that, if impure materials were used, the crystals grown from benzine were very much more impure than were those grown from acetone or carbon tetrachloride. This is another reason for preferring acetone or carbon tetrachloride to benzine, except in cases where crystals elongated along c are needed.

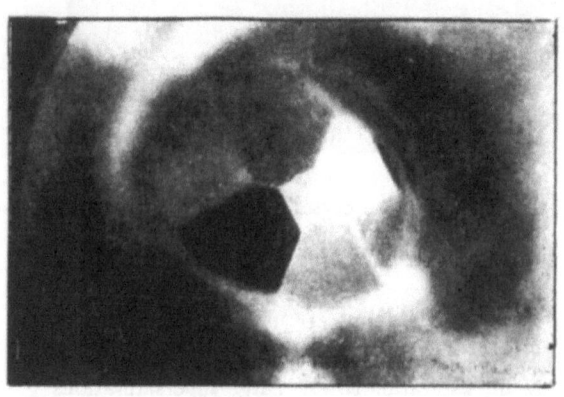

Fig. 5. A negative crystal of benzophenone. × 30.

The {110} prism faces take up less impurities than the others. They also grow the least rapidly.

The {021} faces grow the most quickly, and are always much smaller than the others; they are the first to become cloudy when the supercooling exceeds 0.1°C.

We found that the growth rates of the faces depended on the stirring rate as well as on the supersaturation. No exact relationship could be established. The {110} faces usually grew at constant rates when the growth conditions were constant. The same is roughly true of the normal growth rates of {101} and {111} faces. Sometimes {010} faces appeared on small crystals, but they soon vanished.

The growth rates of {011}, {012}, and {001} faces changed greatly, even within a single experiment. We often found that during a single run (which might last up to 20 days) these faces would appear, vanish, and appear again, sometimes slowly, sometimes suddenly. Nevertheless, the growth rate of the crystal as a whole was so nearly constant under fixed conditions that crystals grown over identical periods (of 10-20 days) were very similar both in form and in size.

We did a few tests in which the growing crystal was turned very rapidly. In these tests the holder was driven by a special motor fixed to the lid. The range of speeds extended up to 8000 rpm. The crystals were transparent even at supercoolings of 0.2-0.3°C. The growth rate of the crystal as a whole was much increased. No {021}, {010} or {001} faces were seen on such crystals, and the {110} prism faces were curved.

Symmetry. The 2:2 morphologic symmetry of the orthorhombic modification has been described by Groth [7] and by others [1].

We have found in the literature no data on the optical activity of benzophenone crystals, and even the piezoelectric data are not reliable [8, 9], and so we undertook to check whether the crystals do in fact fall in class 2:2.

We had available some large perfect monocrystals, which we measured on the goniometer; we also crystallized spheres cut from a benzophenone monocrystal. The results agreed well with the earlier data [1, 7], and so are not given in detail here.

We also observed the etch and impact figures. Negative crystals were well seen in the etch figures on all faces, except for those of the {110} prism.

It was sufficient to place a small (1 mm) crystal of salol on the surface of the benzophenone in order to get an etch figure. In a few hours there formed a hollow filled with a liquid solution of salol in benzophenone. Faces

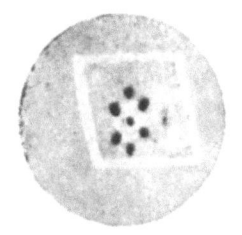

Fig. 6. Positions of the bright spots on the faces of a negative crystal of benzophenone.

are formed in the hollow as the benzophenone liquefies. The process takes 20-40 hrs. The hollow is seen as a negative crystal of good shape (Fig. 5) when the liquid has been removed. We have been able to make these negative crystals in sizes from a few tenths of a millimeter to 3-4 mm. The shapes often correspond well with the ideal growth form of benzophenone.

The positions of the light signals from the faces of a negative crystal were examined by Shubnikov's method [10], and also give data on the symmetry of benzophenone. The results agree well with the 2 : 2 symmetry (Fig. 6).

The fact that benzophenone falls in class 2 : 2 means that two enantiomorphic forms should exist. No data have so far been given on such forms. We have been able to make right-handed and left-handed forms (Fig. 4).

I wish to thank Academician A. V. Shubnikov for guidance in this work, I. S. Zheludev for some valuable discussions, and my colleagues in the Laboratory of Electrical Properties of Crystals for assistance with the experiments.

LITERATURE CITED

[1] R. Nacken, Geol. u. Paläontol. 11, 133-164 (1915).

[2] I. Morris and R. Strickland-Constable, Trans. Faraday Soc. 50, 12, 1378-1393 (1954).

[3] Linnemann, Ann. Chem. u. Pharmacie 133, 1-32 (1865).

[4] M. Kollarits and V. Merz, Berichte d. Deutschen Chemischen Gesellschaft 5, 444-448 (1872).

[5] Pickardt, Z.Phys. Chem. 42, 17-49 (1903).

[6] A. Wells, Phil. Mag. 37, 266, 184-199 (1946).

[7] P. Groth, Chemische Krystallographie 5, 102 (1919).

[8] A. Hettich and A. Schleede, Z. Phys. 50, 249-265 (1928).

[9] S. Elings and P. Terpstra, Z. Krist. 67, 279-284 (1928).

[10] A. Shubnikov, Z. Krist. 78, 111-135 (1931).

SYNTHESIS OF ESPECIALLY PURE CALCIUM AND BARIUM
FLUORIDES FOR GROWING OPTICAL MONOCRYSTALS

I. A. Sinyukova and I. V. Stepanov

The quality of artificial crystals depends very much on the purity of the materials used. In some cases, the commerical compounds are adequately pure. Thus the KCl, NaCl, KBr and certain other salts produced at home and abroad are often suitable for growing optical-grade monocrystals without prior purification. More often, however, the compounds are so impure as to be unsuitable for growing such crystals. The best that can be achieved in such cases is to produce more or less coarse-grained lumps consisting of several solid phases.

Materials that are sparingly soluble in water or the common solvents are usually the most impure. They are substances that cannot be made by precipitation to be pure and free from trapped impurities. The fluorides are compounds of this type. Calcium, barium, and lithium fluorides are among the worst, because they are very sparingly soluble; they are compounds that will be much used in the near future. These compounds precipitate initially as gelatinous colloids (LiF) or as finely divided particles (CaF_2, BaF_2, Fig. 1). The particles grow larger on standing, and gradually settle out. The large specific surface of the precipitate causes much of the impurity to be adsorbed. These adsorbed impurities cannot be removed by chemical treatment, e. g., by acid.

We have found that CaF_2 and BaF_2 may be made as very much purer crystalline powders, with grain sizes of 0.2-0.3 mm. The compounds are made, not as usual by precipitating them from solutions, but by precipitating them from high-temperature melts, e. g.

$$CaCl_2 + 2NaF \rightleftarrows CaF_2 + 2NaCl$$
$$\text{melt} \quad \text{melt} \quad \text{crystal} \quad \text{crystal}$$

Calcium and barium flurodides dissolve in fused sodium chloride and readily give supersaturated solutions. This is easily demonstrated by direct experiment. Sodium chloride, fused in a platinum crucible, dissolves powdered CaF_2 or BaF_2, which compounds separate out as crystalline deposits as the melt cools slowly; they may be highly pure if the saturation and cooling are suitable (Fig. 2).

We worked on a large scale (1.5 kg lots of the components) in our tests. There were three separate types of test: 1) those in which the reactants were used in exactly the amounts required by the equation, 2) those in which the alkaline-earth chloride was in excess, and 3) those in which the sodium fluoride (for CaF_2) or potassium fluoride (for BaF_2) was in excess.

Our methods were as follows. The materials, dried at 250°C, were weighed out, and were mixed in a 1.5 liter platinum crucible. The crucible was closed with a platinum lid and was heated to 1000°C in an electric

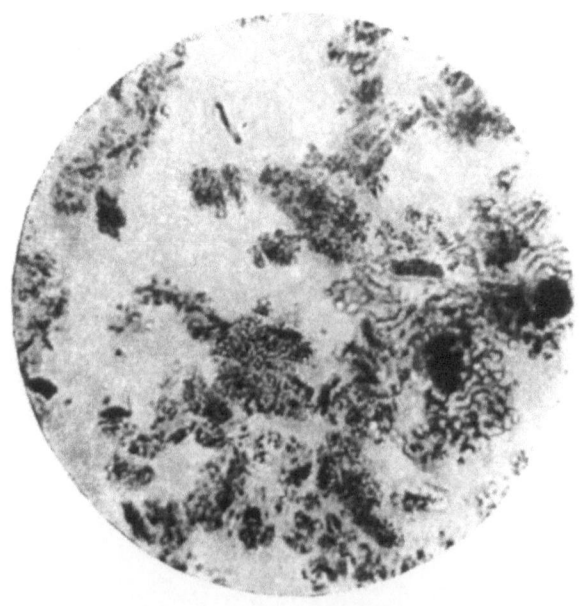

Fig. 1. Finely divided CaF_2 precipitated from aqueous solutions of $CaCl_2$ and NH_4F. The ultimate particles cannot be seen at × 160. Here magnification is × 600.

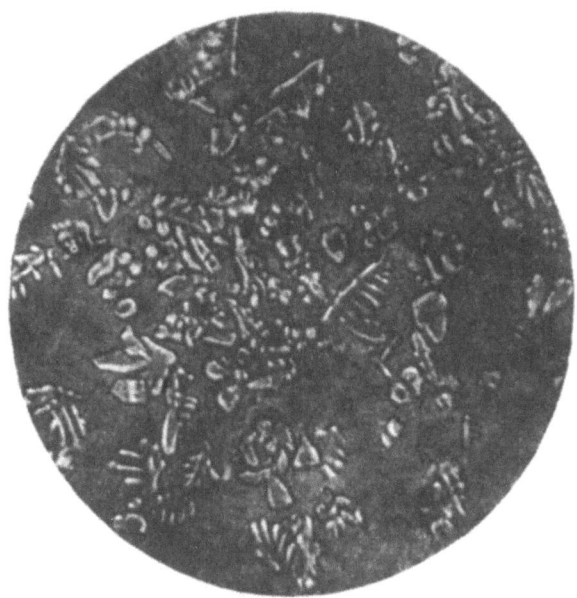

Fig. 2. Crystalline CaF_2 made by dissolving finely divided CaF_2 (Fig. 1) in fused NaCl and cooling the melt. The numerous skeletal crystals occur because the melt was cooled very rapidly. × 600.

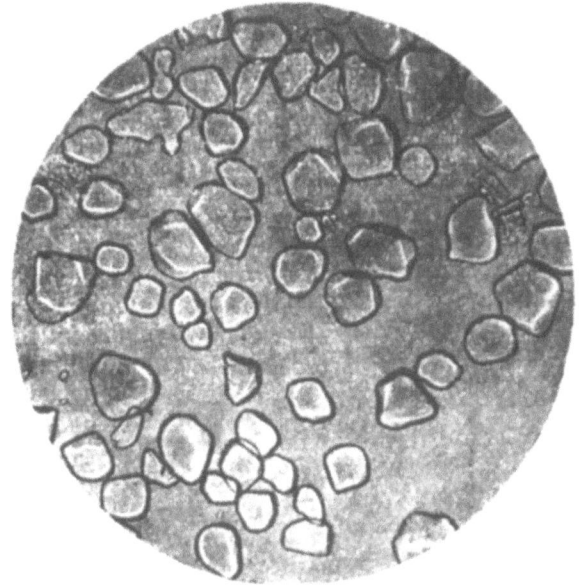

Fig. 3. CaF_2 precipitated from a $CaCl_2$ + NaF melt with excess $CaCl_2$. There are many skeletal crystals with trapped secondary reaction products as well as cubic octahedra. × 600.

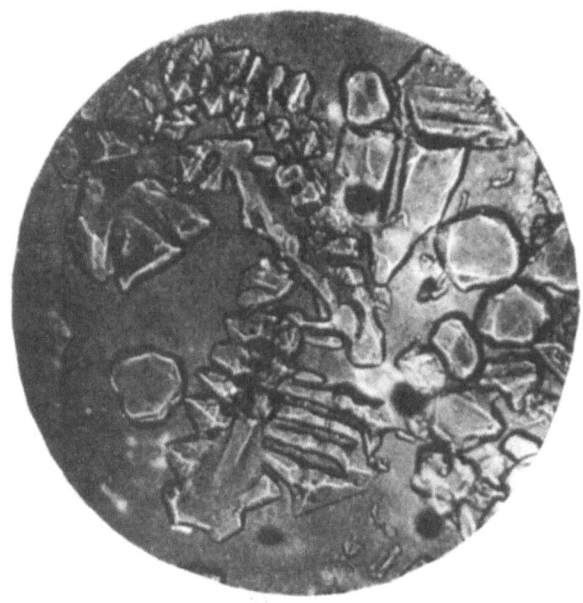

Fig. 4. CaF_2 precipitated from a $CaCl_2$ + NaF melt made in exact accordance with the equation. The rolled appearance of the crystals is caused by the subsequent chemical treatment (prolonged boiling). × 600.

furnace. The charge fused. The charge was kept at 1000°C for three hours and was then cooled slowly (over 4-6 hrs) to 700-750°C, i. e., until it had solidified. It was then left to cool to room temperature, and the NaCl or KCl, which contained CaF_2 or BaF_2, was dissolved out in water containing a little HCl or HF; the residue was then washed with distilled water until the reaction for Cl ion was negative. The washed residue was dried at 2002-05°C in a platinum crucible. The products gave monocrystals that were in every way suitable for optical purposes.

The material was first checked under the mineralogical microscope (at × 200) in a liquid of refractive index 1.43-1.47. This simple method was very convenient, because it could be used also to find the grain size to observe the habit (skeletal crystals, small cubes with octahedron faces, flattened crystal), and to estimate the purity from the proportion of foreign inclusions (which differ in refractive index) in the grains.

We did not analyze the materials, because fluorides are difficult to analyze, and because analytic methods were in adequately sensitive for our purpose. The final test was to grow monocrystals in a vacuum plant [1].

Parts of the crystals that showed inclusions were analyzed spectroscopically (emission and fluorescence spectra). We could not say what the inclusions seen under the microscope were, but we could quickly establish whether the batch of material was suitable for use.

Thus, the CaF_2 precipitated from a melt with excess $CaCl_2$ (twice the theoretical amount) was unsuitable for growing monocrystals. The resulting crystals were usually cloudy. We found that the material contained a birefringent substance between the branches of the skeletal crystals, and many birefringent plates of unknown composition (Fig. 3).

We found mainly skeletal growths and rounded convex grains, rather than regular cubic crystals, when the sodium fluoride was in excess.

Melts whose compositions were exactly in accordance with the equation gave the best results. The products consisted of completely pure cubic octahedra with rounded edges. The grains were 0.1-0.2 mm across (Fig. 4). The best monocrystals are grown only from homogeneous materials.

É. G. Chernevskaya has made strontium fluoride in the same way, and has grown optical-grade monocrystals from it.

Thus, we have confirmed what is well known, but what some have, unfortunately, forgotten, namely that crystalline materials are the only ones that can be made really pure. We were able to grow optical-grade CaF_2, SrF_2 and BaF_2 monocrystals in large sizes only when we had made calcium and barium fluorides of defined grain size and habit, free from foreign inclusions, as crystals of size not less than 0.1 mm.

LITERATURE CITED

[1] I. V. Stepanov and P. P. Feofilov, Growth of Crystals. I [in Russian] (Izd. AN SSSR, 1957) p. 229.

THE EFFECTS OF COOLING CONDITIONS ON THE FORMATION
OF DISLOCATIONS IN GERMANIUM CRYSTALS

E. Yu. Kokorish

Recent progress in semiconductor electronics has required purer and more perfect semiconducting crystals for use in many semiconductor devices.

Germanium is very much used. Methods of making extremely pure germanium have been developed, and attempts are being made to grow germanium monocrystals that in structural perfection approach the ideal limit.

There are several papers on the causes of imperfections in pulled crystals. Thus Billig [1, 2] and Dorendorf [3] have examined the conditions that give rise to dislocations in germanium crystals grown from the melt. It was shown that the causes of the dislocations and the growth conditions are related. Kurtz et al. [4] have examined the effects of growth rate on the perfection of germanium monocrystals grown by zone fusion. They found that the dislocation density increases rapidly at growth rates above 4 mm/min. Indenbom [5] has given a macroscopic theory of the origin of dislocations during crystal growth. He considers that the dislocations form on account of the uneven temperature distribution in the growing crystal.

The macroscopic dislocation density β caused by the temperature field T in the crystal is specified by the tensor

$$\hat{\beta} = -\operatorname{grad} T \times \hat{\alpha}, \qquad (1)$$

where $\hat{\alpha}$ is the thermal expansion tensor for the crystal.

The experiments. We grew several germanium monocrystals of inherent conductivity by pulling from the melt in order to study the distribution of dislocations along the length. The seeds had the [111] orientation, the pulling rate was 1 mm/min, and the crystal was turned at 60 rpm. Wafers were cut at right angles to the growth axis at various distances along the crystal.

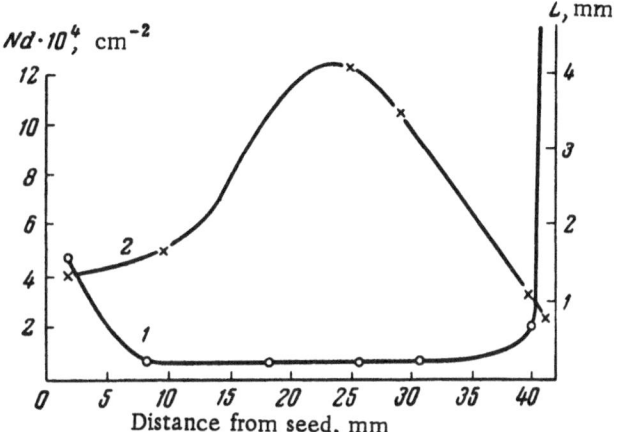

Fig. 1. Dislocation density (1) curve for the monocrystals grown by pulling from the melt.

Fig. 2. Ray-like distribution of etch pits along the [211] directions in a cross section of the neck of a crystal. × 4.5.

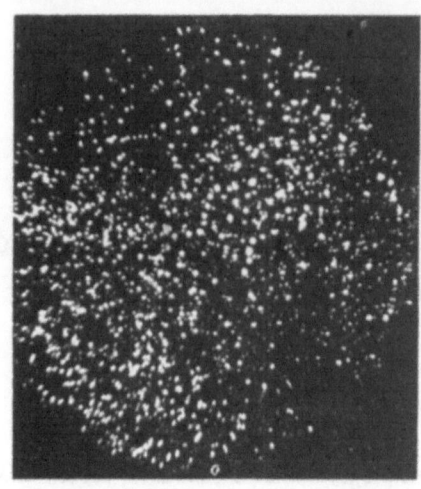

Fig. 3. Random distribution of etch pits in the central part of a crystal. × 4.5.

Fig. 4. Distribution of etch pits at the end of a crystal. × 4.5.

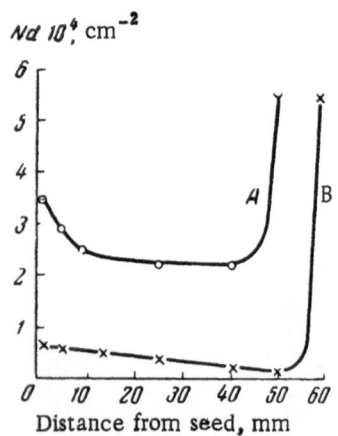

Fig. 5. Distribution of etch pits along (A) crystals grown with forced cooling, and (B) without forced cooling.

The wafers were prepared as follows. They were first ground on glass with M10 and M14 powders in suspension until layers 70-80 μ thick had been removed. They were then washed carefully in distilled water. Next, they were polished for 10-15 sec in a hot acid mixture, $HF : HNO_3 : H_2O = 5 : 5 : 1$. They were then again washed in distilled water. Then followed etching in a boiling solution that contained 12 g of KOH and 8 g of $K_3[Fe(CN)_6]$ in 100 cc of water, followed by a further wash in distilled water and drying.

The pits were viewed and counted with an MIM-6 metallographic microscope (at × 147 and × 280).

The general etch pattern was viewed with an MBS-2 microscope.

Figure 1 shows the etch-pit density (curve 1) and the diffusion length (curve 2). The specific resistance varied evenly from 55 to 50 ohm-cm along the length. Figure 1 shows that the etch-pit density starts off large, but falls to a minimum at the center of the crystal. The density rises sharply toward the end, and is greatest at the very end. Correspondingly, the diffusion length is largest at the center, where the density is least.

The neck in some cases showed a ray-like etch-pit distribution (Fig. 2). Dorendorf [3] states that the etch pits are concentrated in the [211] directions here; this is in fact so. This pattern was not found in all crystals.

The etch pits were usually randomly distributed in the center of a crystal (Fig. 3). The pattern was usually that of Fig. 4 at the very end. The etch pits are concentrated along lines parallel to the three [110] directions, between which lines there are few pits. This pattern is first seen at 25-35 mm from the seed.

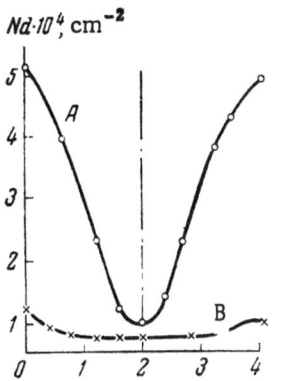

Fig. 6. Etch-pit distributions in cross sectional views for the crystals of Fig. 5.

We examined the effect of cooling rate by cooling come crystals vigorously while they were being pulled. The crystals were cooled at about 15 mm from the surface of the melt by a cooler through which water flowed. The cooler had an internal diameter of 25 mm. The pulling rate was 1.5 mm/min. Figure 5 shows the etch patterns found with these crystals, and with ones grown normally. Crystal A was grown in the above way, and crystal B under the usual conditions. It is clear that the forcibly cooled crystal shows at least ten times as many dislocations as does a normal monocrystal.

The etch-pit density becomes much larger at the end of the crystal because the crystal is lifted out of the melt at this point. The cooling is then so rapid as to be equivalent to quenching. Figure 6 shows etch-pit patterns

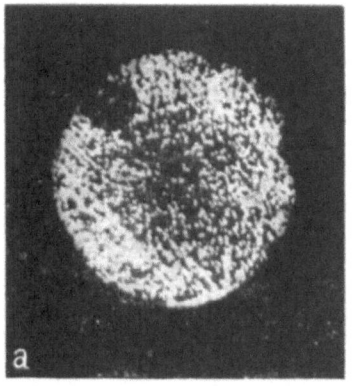

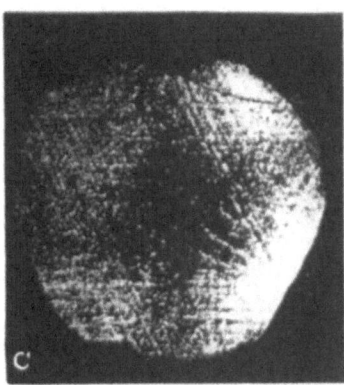

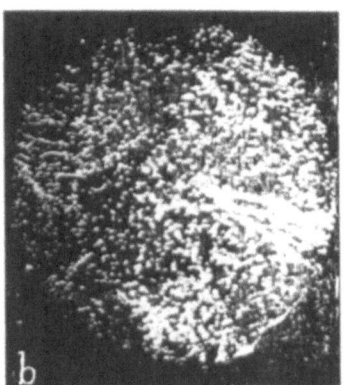

Fig. 7. General etch-pit patterns for various parts of a crystal A. ×4.5. a) 1 mm, b) 25 mm, c) 50 mm from the seed.

for crystals A and B. The sections were cut at 15 mm from the starting end. The forcibly cooled crystal has the most pits at the outside and much fewer in the middle, whereas a normal crystal shows no such marked difference.

Figure 7, a, b, c show etch-pit patterns for different parts of crystal A, while Fig. 8, a, b does the same for crystal B.

Discussion. We have studied the dislocations produced under various growth conditions as the crystals cool from the melting point to room temperature. The dislocations are produced mainly by the thermal stresses caused by uneven cooling in the range in which plastic deformation occurs.

The dislocations are most common at the starting end because the axial temperature gradient in the crystal is then large, since one end of the seed is cooled by running water. At this end of the crystal the diameter is in-

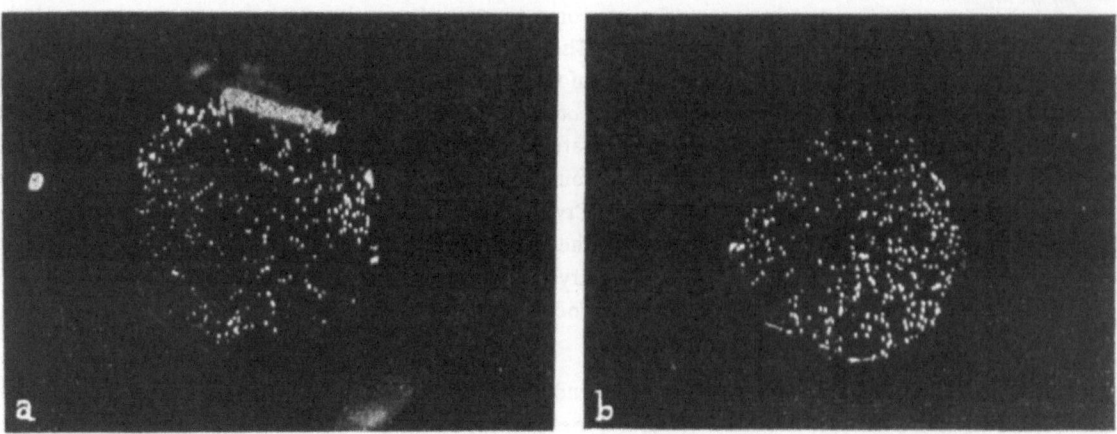

Fig. 8. General etch-pit patterns for various parts of a crystal B. × 4.5: a) 1 mm, b) 25 mm from the seed.

creasing rapidly, and most of the heat is lost through the seed; the isotherms are curved. The middle of the crystal is cooler than the outside at the same distance along the axis, so there is a radial temperature gradient. The combination of these gradients increases the dislocation density.

The axial heat flow falls as the crystal grows; more heat is lost through the sides. The axial and radial gradients are reduced, and the isotherms become nearly flat; hence, the dislocation density is lower in the middle of the crystal.

The density suddenly becomes larger at the end because the crystal is lifted up and taken into the cold zone at this point. The crystal then cools rapidly. The axial and radial gradients become large, and so the dislocation density is high.

The extra forced cooling increases the etch-pit density very much, relative to normal. The density increases because the radial gradient is large, as Fig. 6 shows indirectly; the center and the periphery show very different densities. The axial gradient increases also, and so the etch-pit density is everywhere higher than normal. Crystals with high dislocation densities caused by forced cooling have poor electrical properties, especially as regards minority-carrier lifetime.

I should like to thank Dr. N. N. Sheftal' for guidance in this work, and V. K. Bichev for assistance with the experiments.

LITERATURE CITED

[1] E. Billig, Proc. Royal Soc. 235, 37 (1956).

[2] E. Billig, Brit. J. Appl. Phys. 7, 10 (1956).

[3] H. Dorendorf, Z. angew. Phys. Einschl. Nukleonik 9, 513 (1957).

[4] A. Kurtz, S. Kulin, and B. Averbach, J. Appl. Phys. 27, 1287 (1956).

[5] V. L. Indenbom, Kristallografiya 2, 594 (1957).

GUANIDINE ALUMINUM SULFATE HEXAHYDRATE

SYNTHESIS, PROPERTIES, AND MONOCRYSTAL GROWTH

(Preliminary Communication)

I. S. Rez and L. A. Varfolomeeva

The new ferroelectric guanidine aluminum sulfate hexahydrate (GAS) $[C(NH_2)_3] Al(SO_4)_2 \cdot 6H_2O$ was first made in 1955 by Holden et al. [1, 2]; Anderson's data [3] show that it is a promising material for use in the stores of digital computers.

The published data on GAS are as follows. Ferraboschi [4] was the first to make the substance (in 1908) as small crystals; he described the habit and etch figures, and measured the refractive indices as n = 1.5423 (1.45540), and the density as ρ = 1.806. The crystals are short hexagonal prisms ending in basal planes. No other faces have been observed. Sharp etch figures appear if the basal faces are placed briefly in water; they are flat triangular pyramids with equilateral bases, with one edge parallel to a prism edge. The crystals cleave perfectly on the basal planes. These data have been confirmed by later work in the USA [5], in the TsNILP,* and in the Institute of Crystallography, Academy of Sciences of the USSR.

Goniometric data [4] show that GAS falls in the trigonal system, and x-ray data [12] give the space group as C_{3V}^2 — P31m, with three molecules in the unit cell.

In 1925 Canneri [6] grew double salts isomorphous with GAS of the general formula

$$[C(NH_2)_3] \, Me^{3+} (SO_4 \cdot 6 H_2O),$$

where $Me^{3+} = Cr^{3+}, Fe^{3+}, V^{3+}$.

Holden et al. [5] have examined the chromium and gallium members of this series, and also the double selenates of guanidine with aluminum, gallium and chromium. They found that the $[C(NH_2)_3]^+$ ion could not be replaced by carbamide, thiocarbamide, methylguanidine, aminoguanidine, formamidine, or acetamidine, which do not have the trigonal symmetry, or by trimethylsulfonium ion, which (unlike the guanidonium ion) is not planar, and cannot form hydrogen bonds.

It may be that N—H—O bonds have a larger role than is usually assumed in causing ferroelectric behavior in such structures.

Latterly Matthias and Remeika [7] have detected a low-temperature ferroelectric transition in $(NH_4)_2SO_4$, and Jona and Pepinsky [8] have found an analogous transition in the isomorphous langbeinite, $(NH_4)_2Cd(SO_4)_3$.

Pepinsky et al. [9] have also detected low-temperature ferroelectric transitions in the alum-type compounds $R^+Me^{3+}(SO_4)_2 \cdot 12H_2O$, where R is $[C(NH_2)_3]^+$ (guanidine), $CH_3NH_3^+$ (methylammonium), $CO(NH_2)_2^+$ (carbamide), etc.

This all shows that GAS is merely one representative of an extensive class of new ferroelectrics, all of which have hydrogen bonds and complicated structures, and most of which are hydrates. The compounds are all of complex type, which shows that we are much more likely to find new piezo- and ferroelectrics in selected families of hydrated complex compounds than we are in other classes of compound.

---

* Central Scientific-Research Laboratory for Fruits and Vegetables.

GAS is ferroelectric from 200°C (at which temperature it soon decomposes) down to liquid-nitrogen temperature [5]. The L_3 axis is the ferroelectric one. The hysteresis loop at 60 cps is often inclined or has kinks. The shape of the loop depends on the orientation of the crystal. At room temperature the spontaneous polarization is about 0.35 μ/cm^2, and the coercive force at 60 cps is 1000-3000 v/cm. These values increase as the temperature falls. The permittivity in weak fields is about 6 along L_3, and about 5 normal to that axis. The isomorphous compounds have roughly the same ferroelectric constants as GAS.

Synthesis

No data are given on the synthesis in the literature, so a synthesis had to be developed at TsNILP. One of the present authors did the first synthesis, from guanidine and aluminum sulfates (both chemically pure). Concentrated equimolar solutions of the components were mixed at room temperature, and the mixture was evaporated for 1.5-2 hrs. The crystalline deposit was recrystallized twice, and the first small monocrystals of good grade were grown.

Subsequently we used a rather more complicated two-step synthesis from the readily available guanidine carbonate. This carbonate (pure) was recrystallized twice, after which sulfuric acid (chemically pure) was added with stirring to the solution until no more CO_2 was evolved. The calculated amount of aluminum sulfate (chemically pure) was then added to the solution. The solution was then evaporated for 1-2 hrs, was filtered hot, and was left to crystallize. The GAS which appeared was recrystallized 3 or 4 times, until the solutions were quite colorless and did not become cloudy on prolonged heating to 70°C. The latter condition is essential to growing good monocrystals. The solution tends to become cloudy if the GAS is overheated at any stage, especially when the crystalline raw material is being dried. All drying should therefore be done below 70°C.

We also made the isomorphous chromium compound $[C(NH_2)]_3Cr(SO_4)_2 \cdot 6H_2O$ (GCS), for use in x-ray studies. As with GAS, we made the compound by mixing saturated solutions of the components (guanidine sulfate and chromic sulfate) at room temperature. Now $Cr_2(SO_4)_6 \cdot 18H_2O$ is very readily hydrolyzed, and the violet form is transformed to the green in solution at +60°C (which form does not give GCS), so the solution was evaporated without heating in a drying cabinet. We obtained characteristic hexagonal tablets of GCS, which were examined with x-rays.

The Solubility of GAS

We measured the solubility of GAS in water over the range from 22.7 to 88.2°C by Alekseev's method (modified), in which a U-8 flow thermostat is used [11]. Table 1 gives the results.

TABLE 1

Solubility of GAS in Water

Temperature, °C	Solubility, g/solubility	Temperature, °C	Solubility, g/solubility
22.7	592	75.6	901.45
26.1	637.7	79	940
28.2	686.2	81.7	990.8
42.9	728.9	86	1044
64.45	812.5	88.2	1088.5
70.2	864.2		

The crystals were dehydrated to a total loss in weight of 27%, which corresponds to 5.7 of the 6 molecules of water; the hydrate is almost completely dehydrated by heating at 0.5°C/min over the range from 100 to 200°C.

Growing GAS Monocrystals

We grew the monocrystals under laboratory conditions from point seeds by the isothermal static method. The supersaturation was controlled in terms of the evaporation rate. The crystallizers were conical flasks set in a thermostatted water bath, whose temperature was kept constant to ± 0.1°C. The optimum temperature was 40-45°C.

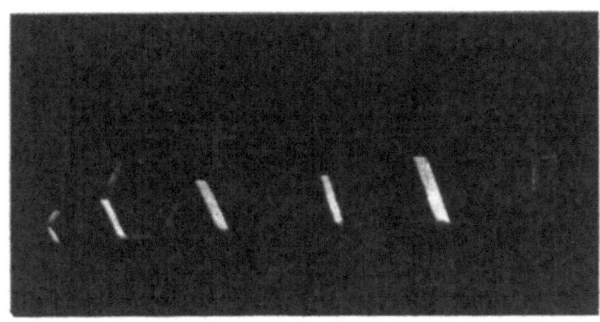

Fig. 1. Crystals of guanidine aluminum sulfate hexahydrate.

The evaporation rate was adjusted by varying the permeability to water vapor of a stack of filter papers in the neck of the flask; the number of layers was 4-5 at the start, and was reduced to two by the end. We also used interchangeable top layers with differing areas lacquered. The growth rates were 1.5-2 mm/day along X and Y, and 0.5-0.7 mm/day along Z. The finished crystals weighed up to 250 g.

The solution is very sensitive to impurities. The hindered growth remarked on by Chapelle et al. [15] we found only if contaminated solutions were used. The results were very reproducible if pure solutions were used and the supersaturation was controlled correctly.

Figure 1 shows some of the crystals. The circular defects at the centers reproduce the slightly convex bottom surfaces of the flasks.

X-ray Studies on GAS and GCS Monocrystals

The fact that isomorphous series of general formula $[C(NH_2)_3]Me(\partial O_4)_2 \cdot 6H_2O$, where Me = Al, Cr, Ga or V, and ∂ = S or Se, exist [1, 5, 10] was of great assistance in elucidating the structure of GAS. Table 2 gives the published density [4] and x-ray [12] data for GAS.

TABLE 2

Crystal	System	Diffraction symmetry	a_0, A	C_0, A	Z	ρ pyc	Space group
Gas	Trigonal	$D_{3d} - \bar{3}m$	11.77 ± 0.04	8.98 ± 0.03	3	1.806^* $1.799 \pm$ 0.005^{**}	$C_{3v}^2 - P31m$

* From [4].
** TsNILP data.

GAS and GCS monocrystals were used. Goniometry, measurements of physical properties, and an x-ray method [13] were used. Table 3 gives the results.

TABLE 3

Crystal	Bravais lattice	Cell parameters, A	$v, \mathring{A}^3$	z	ρ roentgen	Space group
GAS	Primitive	$a = 11.73_{7\pm2}, C = 8.94_{8\pm1}$	1067.4	3	1.818	$C_{3v}^3 - P31\,m$
GCS	"	$a = 11.75_{9\pm2}, C = 9.03_{6\pm2}$	1082.2	3	1.909 $1.885 \pm 0.005^*$	"

* ρ_{pyc} (TsNILP).

The unit cell parameters were measured very exactly by using reflections with the angles of reflection in the 50-85° range. Oscillation photographs were recorded for every reflection with not less than two wavelengths. The radiations were CuK_{α}, Mo and W.

The patterns recorded for the a and c axes showed that there were no characteristic extinctions, and so the diffraction group is $\bar{3}mc$. This group contains six space groups:

$$D_{3d}^1 - C\,\overline{3}1m, \quad D_3^1 - C\,312, \quad C_{3v}^1 - C\,3m1,$$

$$D_{3d}^3 - C\,\overline{3}m1, \quad D_3^2 - C\,321, \quad C_{3v}^2 - C\,31m.$$

Now D_{3d}^1 and D_{3d}^3 drop out, because they have centers of symmetry, whereas the crystals are piezoelectric. Classes 3:2 and 3·m can be differentiated from the optical activity. This activity was such as to be unmeasurable with GAS. Hence 3:2 drops out. The group $C_{3v}^2 - C31m$, alone of the two remaining, agrees with the choice of crystallographic axes and with the symmetry of the Laue pattern.

The patterns recorded for identical directions in GAS and GCS crystals showed only slight changes in the intensities.

The structure of GAS may be described in terms of the formula $[C(NH_2)_3]^+[Al(H_2O)_6]^{+3}(SO_4)_2^{-2}$. The molecule consists of four structural units, namely the $[C(NH_2)_3]^+$ triangle, the $[Al(H_2O)_6]^{+3}$ octahedron, and two $(SO_4)^{-2}$ tetrahedra. The structure resembles that of alum, in which the molecule has two octahedra and two tetrahedra.

SUMMARY

Laboratory methods of synthesizing GAS and GCS have been developed and GAS monocrystals have been grown. These latter have had weights up to 250 g; the solubility of GAS in water, and the densities of GAS and GCS, have been measured. The temperature course of the dehydration of GAS has been studied. The cell parameters and space group of GAS and GCS have been established, and a start has been made on detailing the structure of GAS.

We should like to thank L. Z. Rusakov, G. S. Zhdanov, M. M. Umanskii, and A. A. Shternberg for discussions, and Yu. A. Belyakova and E. E. Tsepelevich for assistance with the experimental work on the syntheses and on growing the GAS monocrystals.

LITERATURE CITED

[1] Holden, Matthias, Merz and Remeika, Phys. Rev. 98, 2, 546 (1955).

[2] Holden, Matthias, Merz and Remeika, Bull. Amer. Phys. Soc. 30, 3 (1955).

[3] Anderson, Bell Lab. Record No. 5 (1955).

[4] Ferraboschi, Proc. Cambridge Phil. Soc. 14, 5, 471-474 (1908).

[5] Holden, Matthias, Merz and Remeika, Phys. Rev. 101, 3, 962-966 (1956).

[6] G. Canneri, Gazz. Chim. Ital. 55, 611-615 (1925).

[7] Matthias and Remeika, Phys. Rev. 103, 1, 262 (1956).

[8] Jona and Pepinsky, Phys. Rev. 103, 4, 1126 (1956).

[9] Pepinsky, Jona and Shirane, Phys. Rev. 102, 4, 1181 (1956).

[10] Remeika and Merz, Phys. Rev. 102, 1, 294 (1956).

[11] N. S. Rez and L. I. Tsinober, Growth of Crystals. I [in Russian] (Izd. AN SSSR, 1957).

[12] E. Wood, Acta Cryst. 9, 17, 618 (1956).

[13] G. S. Zhdanov, M. M. Umanskii, L. A. Varfolomeeva, Z. I. Ezhkova and Z. K. Zolina, Kristallografiya 1, 3 (1956).

[14] N. S. Rez, Izv. Akad. Nauk SSSR, Ser. Fiz. 21, 3, 466-472 (1957).

[15] J. Chapelle and J. Chollof, Compt. rend. 244, 5, 1185-1187 (1957).

THE CRYSTALLIZATION OF GERMANIUM ON SILICON AND
OF SILICON ON GERMANIUM

N.P. Kokorish

Germanium and silicon both have diamond-type lattices, and their lattice parameters differ by only 4.03% (a_{Ge} = 5.65 A, a_{Si} = 5.42 A, a_{Ge}/a_{Si} = 1.0402).

Stöhr and Klemm [1] state that silicon and germanium form continuous and substitutional solid solutions. There is, however, much intracrystalline liquation when a germanium–silicon melt crystallizes, which effect makes it difficult to get homogeneous materials, especially monocrystals.

It is claimed that alloys can be made of germanium with up to 24% Si, and of silicon with up to 6% Ge [2]. These alloys, on the germanium side, are of interest in semiconductor technology. They have forbidden bands wider than that in germanium, and so the working temperature range can be extended. They also have higher carrier mobilities than silicon does.

Hence, the crystallization of germanium on silicon and of silicon on germanium is of some interest. An overgrowth of one on a monocrystal plate of the other may be of practical value.

Germanium on silicon. The monocrystal plates were 0.3-1 cm^2 in area and 0.4-1 mm thick. These wafers were cut mostly on (111), accurate to ± 2°. The wafers were ground and polished, and were then etched for 2-4 min in boiling 30% NaOH. Figure 1 a and b, shows wafers etched for 1 and 3 min. Characteristic triangular etch figures are seen. The corners of the triangles are rounded, and many rows of steps run parallel to the sides. The triangles became larger and less regular as the etching time was increased. The wafers were dried at 100°C for 30 min, and were then placed on a graphite holder in a reaction tube, in which germanium was produced from GeCl$_4$ vapor by the hydrogen method [3].

The reaction oven was used at temperatures of 750 to 940°C for this purpose. Below 800°C the germanium would not grow (none appeared on the silicon wafers). It began to grow at 850°C and continued to do so up to the melting point of germanium (958°C).

We found that the germanium would sometimes not deposit on one or two silicon wafers even above 850°C, although it deposited perfectly well on adjacent wafers.

Figure 2 shows the surface of a wafer after 15 min. The germanium is seen as black spots; the layer is not continuous. Deposition occurs preferentially on the faces of steps formed by the etching, and reveals the surface relief; the appearance resembles that produced by dewdrop or ammonium chloride methods of revealing etch features.

Figure 3 shows the same surface at a higher magnification. The germanium is seen as triangular crystals. These lie parallel to one another. Where they are many, they fuse together and form a layer with gaps, through which the surface of the wafer is seen.

Figure 4 shows the same wafer after germanium had been deposited for an extra 40 min. The layer is still not continuous, but the covered areas are larger.

Figure 5a shows germanium on a natural (111) face of silicon (deposition time 10 min). Here again the layer is patchy. The effect is well seen at high magnification (Fig. 5b). The parallel array of crystals is also apparent.

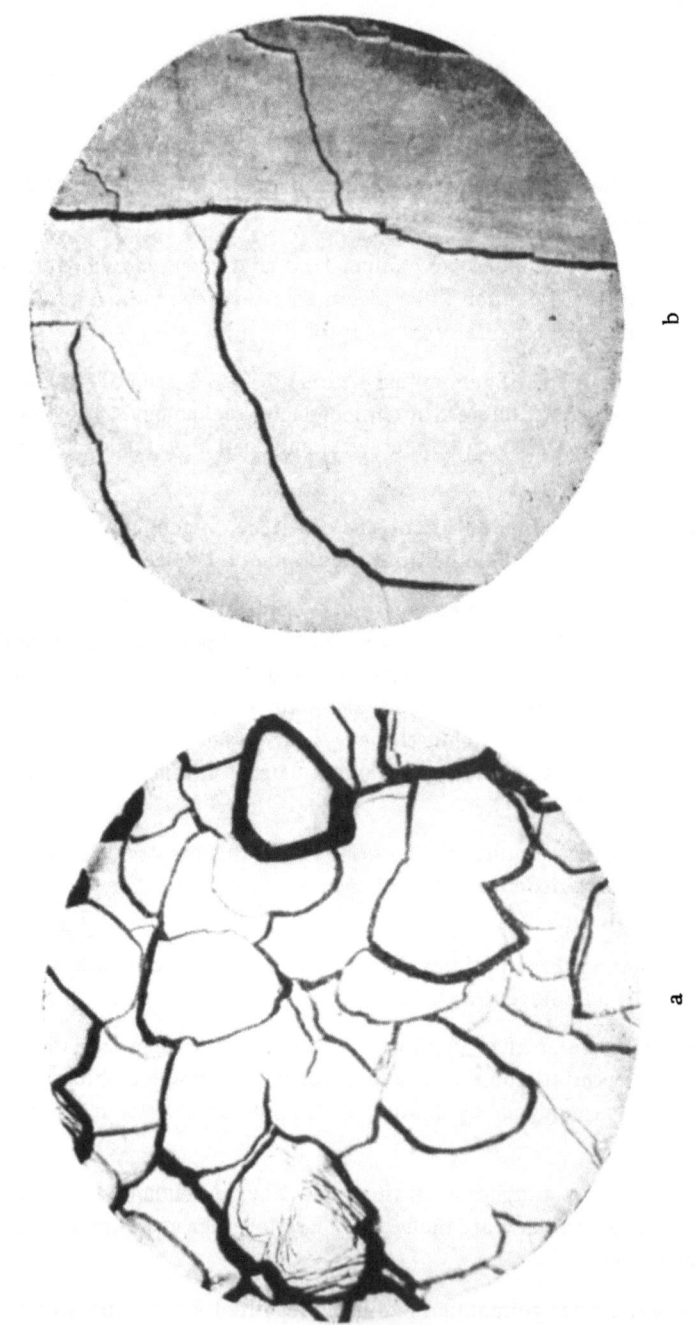

b

a

Fig. 1. Etched surfaces of monocrystal silicon wafers cut on (111): a) 1 min; b) 3 min. × 380.

Fig. 2. Germanium on a monocrystal silicon wafer cut on (111). × 65.

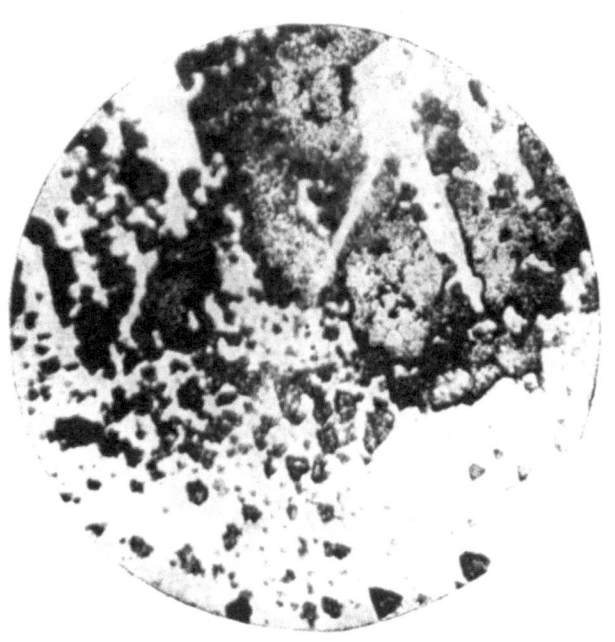

Fig. 3. Same as Fig. 2, but × 380.

We used also silicon wafers cut parallel to (100) and (110). Continuous ordered layers were produced on (110). These layers could be made 80 or 100 μ thick, but they adhered badly, and flaked off when the wafers were rubbed.

Thus, germanium will form an oriented overgrowth on silicon at 850°C or above, but the layers are not continuous, or strong.

Silicon on germanium. We were able to deposit silicon on germanium only within a very narrow temperature range, because hydrogen starts to attack SiCl$_4$ appreciably only at 880°C, and germanium begins to melt at oven temperatures over 940°C.

Fig. 4. Same as Fig. 2, but after extra germanium had been despoited for
40 min. × 65.

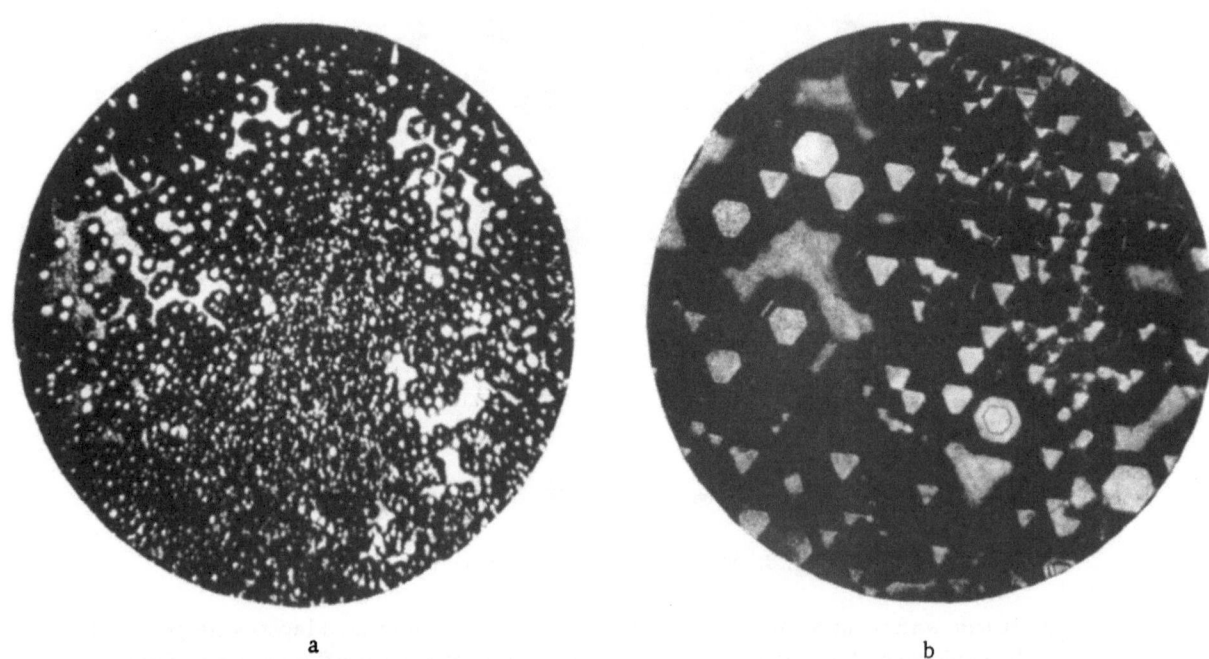

a   b

Fig. 5. Germanium growth on a natural (111) face of silicon in 10 min. The oriented nature of
the overgrowth is clear: a) × 65, b) × 380.

The monocrystal germanium wafers were cut parallel to (111), accurate to ± 2°. The wafers were ground,
were polished with chromic oxide, and were then etched in H_2O_2 for 10 min. The dried wafers were placed in the
reaction tube, and silicon was produced by reducing the tetrachloride with hydrogen.

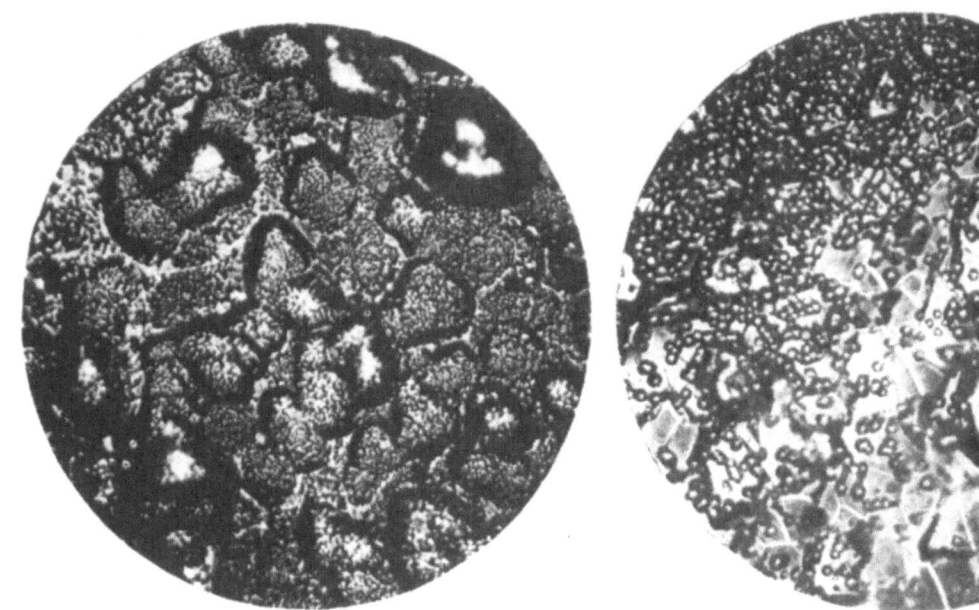

Fig. 6. Silicon on a monocrystal germanium wafer cut parallel to (111). × 900.

Fig. 7. Part of a monocrystal germanium wafer not fully covered by silicon. × 900.

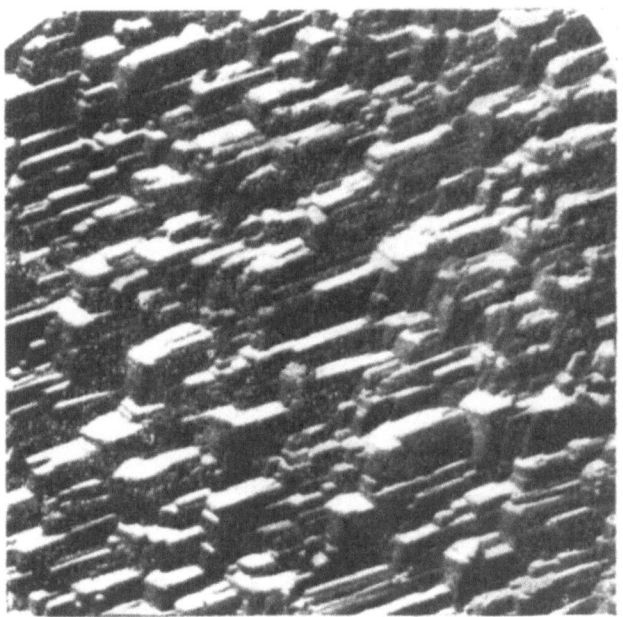

Fig. 8. Electron micrograph of a silicon crystal coated with oriented germanium. × 10,500.

Figure 6 shows a specimen on which silicon had been deposited at 900-920°C. The mean thickness of the layer (from its weight) was 1 μ. The coating was continuous, except in a few places. One such place is shown in Fig. 7; it is clear that the silicon forms triangular crystals that are parallel to one another (oriented overgrowth).

The coated wafers were studied in the electron microscope. Figure 8 shows a continuous layer, formed from parallel silicon crystals that have fused.

The best overgrowths were produced at 940°C, i. e., near the melting point of germanium. The layers were light in color. At × 380, the triangular fused crystals are seen clearly. All the films were continuous.

At 900°C the layers were dark in color, and the picture seen at the same magnification was unclear, and no triangular crystals could be distinguished.

Thus, we could not produce good oriented overgrowths of germanium on silicon, but did get good samples of silicon on germanium.

This is a preliminary communication.

I should like to thank Dr. N. N. Sheftal' for guidance in this work, and M. V. Gavrilova for assistance with the experimental work.

LITERATURE CITED

[1] Stöhr and Klemm, Z. anorg. Chem. 241, 405 (1939).

[2] Hebert, Proc. IRE 11 (1957).

[3] N. N. Sheftal', N. P. Nokorish, and A. V. Krasilov, Izv. Akad. Nauk SSSR, Ser. Fiz. 21, 1, 147-152 (1957).

Institute of Crystallography, Academy of Sciences, USSR

THE GROWTH AND USES OF GEM-GRADE CORUNDUM CRYSTALS*

S. K. Popov

Foreword from the Compiler

The author of this work, Savva Kirillovich (Savelii Keropovich) Popov, 1900-1953, began to work in 1932 on apparatus for producing synthetic corundum industrially. From 1935 until his death he did this work in the Institute of Crystallography, Academy of Sciences of the USSR, and headed the corundum synthesis group, under the general supervision of Academician A. V. Shubnikov.

In 1938, Popov produced a new and semiautomatic device for growing corundum, which device was used to start the production of corundum crystals in the USSR. He made corundum monocrystals as long thin rods at that time. These rods were much easier to use than massive boules for making watch parts, and were much more economical in diamonds.

S. K. Popov

---

* Prepared for publication from Popov's manuscripts and papers.

The second stage (1945-1950) was to develop and introduce an improved device for growing ruby rod, and to set up a production battery of such devices. Difficult problems had to be solved, which were concerned with supplying the charge very accurately and with heating the crystals evenly. The first article deals with this topic.

At the end of his life (1951-1953), Popov developed a new use for rod corundum, namely as a fiber-carrier for use in making artificial fibers. To this end he developed a method of flame-polishing corundum in which diamonds were not used, a method of bending corundum rods, and equipment to do these operations. The second article deals with this work, which was done in collaboration with A. A. Popova.

These articles represent Popov's original work, and contain no additions other than those in the footnotes. They contain only his own personal results. Hence, aspects with which Popov himself did not deal with are either not dealt with at all, or else are dealt with very briefly. Thus, there are only a few lines on the charge in the first article.

The first article has been compiled from materials left by Popov; it deals with his apparatus for synthesizing corundum, and with the technique of growing corundum boules and rods. A later improved apparatus for use in especially accurate work is described. Simpler equipment can be used to grow boules, and so in almost all cases the earlier simpler designs are described. The appendix deals with some purely technological matters, which are, however, important to the process. The various forms of the apparatus are analyzed, because the apparatus may be the subject of major improvements, especially in the light of comparisons made with equipment used in other countries. Popov deals somewhat in outline, but correctly, with the relation between the surface shape and uniformity of a growing crystal. I have not thought it desirable to alter this section, because to do so would have meant rewriting it completely, and especially because another publication from the Institute of Crystallography* contains a detailed treatment of this and related topics.

Popov's solution to the problem of making uniform crystals has been to develop devices to provide the optimum feed and heating conditions. The equipment used in Western Europe would appear to be much more primitive, and the crystals somewhat less good; there, more attention has been given to chemical aspects of the process. It would be of interest to compare Soviet and foreign apparatus and crystals. Examples of foreign products are Czech and German apparatus.

It is hoped that these articles will be found of value to those dealing with synthetic corundum, or with crystal-growing. We hope also that they will preserve the memory of S. K. Popov, a talented designer and experimenter who devoted his life to synthesizing monocrystals.

Z. I. Zhmurova and A. A. Popova assisted in preparing Popov's manuscripts for publication. Sh. S. Popova, Popov's widow, gave us much help. L. M. Shamovskii made some valuable comments on the manuscript.

N. N. Sheftal'

I. GROWING SYNTHETIC CORUNDUM AS RODS AND BOULES

(APPARATUS AND TECHNOLOGY)

INTRODUCTION

Synthetic corundum, α-Al_2O_3, is one of the more important technical monocrystals. It is very hard (9 on Mohs scale), is very resistant to wear, retains its polish against scratching, is chemically stable, and has a small expansion coefficient. These are the properties that make it valuable for bearings, knife-edges and the like. It is used mainly in timepieces (watch stones), in electricity meters, in balances, in meters for electrical and naviga-

---

* V. Ya. Khaimov-Mal'kov and V. A. Perl'shtein, On the Question of the Influence of the Temperature Gradient of the Furnace on the Distribution of Impurities in Crystals. Collected Materials II, Conference on Scintillation (1958).

Fig. 1. A crystal on a refractory rod.

tional purposes, in sound recording, and so on. Latterly, it has come into use for fiber guides in making artificial fibers.

The industrial demand for synthetic corundum may be judged from the daily output of European factories producing watch jewels, which in 1929 was 750,000 carats [1], and in 1942 was 1 million carats (200 kg). The present output is not known, but it must be much larger.

The making of synthetic corundum was considered in the first half of the 19th century. In 1837, Gaudin [2] made the first very small crystals by firing a mixture of potassium sulfide with powdered alumina in a crucible covered with charcoal. In 1851, Senarmont [3] made α-Al_2O_3 as microscopic rhombohedra by heating solutions of aluminum chloride to over 350°C in sealed tubes. In 1848 Ebelman [4] made ruby by fusing alumina and boric acid with chromic oxide. In 1877 Fremy and Feil[5] made crystalline corundum by fusing alumina with lead oxide in a fireclay crucible; the lead aluminate was attacked by the silica in the crucible, and the alumina so liberated crystallized as small plates. The addition of 2-3% of chromium salts made the crystals a ruby-red color. Watch jewels were made from these crystals. In 1887 Fremy and Verneuil [6] made ruby by fusing barium and calcium fluorides with cryolite and Cr_2O_3 at high temperatures. Many others also worked on corundum. Thus, Ebelman and Parmantier [7] dissolved alumina in fused salts that either were volatile at high temperatures or would decompose. The alumina appeared in crystalline form after periods of heating that sometimes lasted for weeks. Gaudin and Moissan [7] made corundum by crystallizing fused alumina.

Some (St. Clair-Deville and Caron, Daubré, Menier, Verneuil [7]) decomposed halides and other compounds of aluminum in steam, with boric acid, and with other reagents that gave aluminum oxide. Finally, compounds containing alumina were decomposed by heat alone (Gaudin, Daubré [7]).

None of the work done up to 1902 was of much practical importance, because the crystals produced were very small.

Verneuil was the first to make corundum crystals of a useful size in 1902. He designed a novel apparatus for growing ruby crystals. In 2-3 hrs the crystals grew to 10-20 carat size (2-4 g); they had rounded shapes (boules). Verneuil's method has been used since then without essential change throughout the world. There have been many technical improvements to the apparatus, which have made it possible to make single crystals of 100-150 carats in 2-3 hrs (Fig. 1).

The general outlines of the method are as follows. Chemically pure alumina is made by firing finely powdered ammonium alum (the particle size is 1-2 μ); this powder is fed continuously in a thin stream through an oxyhydrogen flame (temperature over 2000°C) onto a refractory rod (corundum or silit). A cone of sintered particles builds up on the rod, and then from the top of the cone there grows a single crystal as a thin rod or round boule. The purest alumina is used to give colorless corundum (white sapphires). Metal oxides are added to the alum if colored crystals are required. Red (ruby) crystals are made by adding 1.0-7.0% of chromic oxide, blue ones by adding 1% of titanium oxide and 2% of iron oxide, green ones by adding vanadium and cobalt oxides, and synthetic alexandrites by adding vanadium oxide only.

To make watch jewels, it is best to use uniform crystals free from gas bubbles and cracks. The red variety is preferred for this purpose, partly for esthetic reasons, and partly because the colored stones are less easily lost during manufacture (1000 balance-wheel stones weigh only 1 g).

The description given here relates to the technique of growing corundum monocrystals as bottles and rods, and to an improved apparatus for doing this on an industrial scale.

Crystallization in Boules and Rods. The Crystallization Conditions for Synthetic Corundum, and Their Relation to the Quality of the Product

It has been found that the crystallization conditions and the quality of the crystals are related.

The state of the growth surface is especially important, because this affects the homogeneity of the crystal. The surface may be convex, flat or concave (Fig. 2). In each case the crystal can grow free from inclusions and of true cylindrical shape. The crystals differ in quality, though. The cause is that the shape determines the thickness of the fused layer in which the material gradually crystallizes.

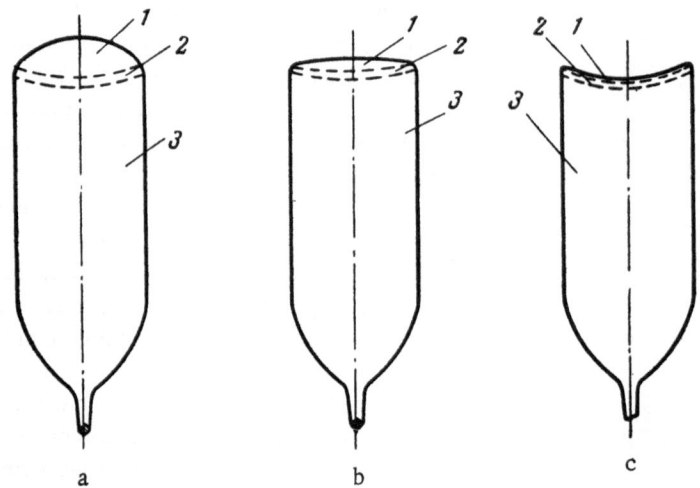

Fig. 2. The growth surface of a crystal: a) spherical, b) flat, c) concave;
1) melt, 2) crystallizing layer, 3) monocrystal.

The tests to be dealt with below have shown that the melt is deepest when the surface is convex, is thinner on flat surfaces, and is thinnest on concave surfaces. The shape of the molten layer, and hence, the shape of the surface, depends on the total heating power as well as the flame temperature.

The thickness of the molten layer was measured by removing it. The reduction in height of the crystal was then measured with a viewing device fixed to the furnace. This viewing device had previously been calibrated with a millimeter scale put in place of the crystal.

Our measurements showed that a light-ruby crystal 20-22 mm in diameter, and having a convex surface, had a fused layer about 1 mm thick. The thickness was 3-5 times smaller if the surface was flat.

The thicker the film of melt, the less the temperature gradient between top and bottom, and, hence, the more gradual and uniform the crystallization process, other things being equal.

Figure 2a shows a convex crystal. The growing part of the boule consists of three zones. The top (convex) one is the melt, and the bottom one, which is cylindrical and has a pointed base, is the monocrystal. Between these two there lies a third zone, the layer which is crystallizing.

A boule that has had a regular convex surface throughout its growth is more uniform, because every layer will have crystallized under the best conditions. The permissible crystallization rate is higher, because the fused film is thicker.

• The quality of the crystal depends on the ratio of the temperature gradient to the feed rate rather than on the gradient alone — Compiler.

Figure 2b shows a crystal with a flat surface. This shape always occurs if the total heat supply is too low, no matter what the flame temperature. The fused film is thinner, the conditions are less good, and the growth rate is lower.

Crystals grown in the latter way always show many more gas bubbles and other flaws. The result is also that the crystal is overstressed, and that hidden cracks may occur. These cracks appear during storage and cutting, either in the crystals themselves, or in the finished parts.

These crystals also show "runs," which are caused by local overheating accompanied by inadequate general heating. This effect occurs if the burner is too small while the flame is extremely hot at the center.

Figure 2c shows a concave crystal. This shape occurs when there is far too little heat and there is excessive pressure on the surface from the oxygen jet. All this causes the fused film to be still thinner. The excessive pressure results from the attempt to compensate for lack of heat by increasing the temperature. The temperature indeed rises, but only at the center; the edge stays solid. The cause of the concavity is thus that the force of the jet pushes the melt to the edges.

Crystals grown in this way are seldom free from inclusions, except when growth is very slow. Their shapes are usually irregular, and the stresses are always large.

Crystals grown under the best conditions have always a thin and even rippled deposit of charge on the sides, which is always a sign of good quality. Mosaic patterns may also be seen on the surface. These patterns denote poor quality.

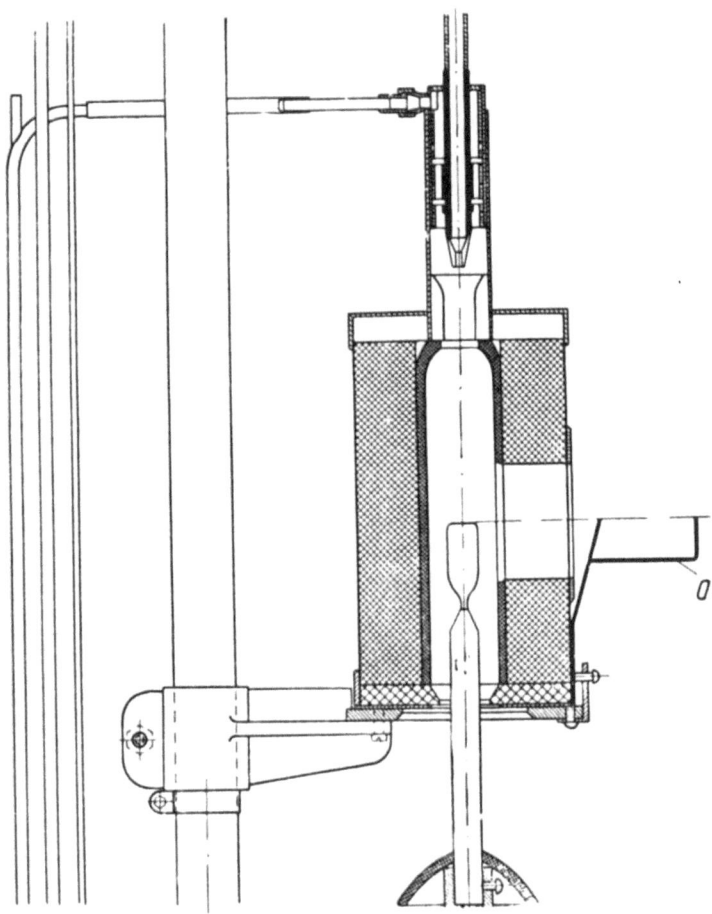

Fig. 3. System for locating the seat of crystallization with a sighting device O.

The temperature must be kept constant on the growth surface throughout the process if the best crystals are to be made. In other words, the surface must be kept at the seat of the crystallization.

This seat of crystallization lies in that zone of the flame which provides the best temperature for the cylindrical part of the crystal to grow. The surface must lie in that zone throughout the growth. The position of this surface is viewed through the viewing device, which is placed outside the viewing port (Fig. 3).

The composition of the gas has to be kept constant in order to keep the temperature constant. It is otherwise not possible to provide a steady heat supply. For the same reason, the gas pressures must be constant.

The surface is kept at the proper point by supply the charge evenly and by withdrawing the crystal at the exact rate at which it grows.

The surface temperature and withdrawal rate are closely related, and an appropriate relation between them must be maintained in order to grow good crystals.

The most difficult part of the whole operation is to ensure that the initial growing point expands properly into a cylinder, because this affects the quality very much. A prepared point or rod seed 2-3 mm in diameter is expanded smoothly (by adjusting the oxygen supply) to 20-24 mm or so. The more evenly the temperature at the surface is raised the better the result. The surface and volume of the crystal increase during this process, while the linear growth rate falls, and so the gas supply has to be adjusted appropriately.

It is most essential to ensure that the region between the seed and the full-sized rod is uniform, because this part of the crystal is the basis on which the rest is built.

The way the grown crystal is cooled also affects the quality very much. The temperature falls rapidly when the gas supply is cut off.

The temperature varies unevenly along and across the crystal. The effect is to increase the stresses which were present even while the crystal was growing. The stresses are also increased if the crystal is withdrawn before it has become cool enough. These residual stresses may crack the crystals, especially in the case of dark ruby ones, which are more sensitive to temperature changes at all stages. The crystals should be cooled very carefully. The viewing port must be closed. The crystal must be withdrawn very slowly. Cold air currents must not fall on the crystal, and the withdrawn crystal should be put in a closed box filled with asbestos wool to cool off completely.

The same treatment must be given to cracked crystals, in order to prevent any further stresses from arising.

Causes of Defects

Changes in temperature, in charge feed rate and in withdrawal speed all affect the crystal.

Figure 4a shows a crystal that has been grown properly, with the best charge feed and withdrawal rates.

Figure 4b shows a crystal with a wrongly formed expanding section.

This expansion can have three causes:

a) the temperature has risen because the calorific value (composition) of the gas has changed;

b) the growth surface has moved up into a hotter area, because the growth rate has exceeded the withdrawal rate;

c) the charge feed-rate has fallen, and so less heat is used up in melting the charge.

In all three cases the charge forms a thinner film on an expanding surface. The linear growth-rate falls, and the growing surface is at a high temperature for a longer time. The cooling conditions are altered, the amount of chromium entering the lattice is affected (because some is burned up), and the crystal becomes lighter in color. The change in composition causes stresses to arise between adjacent layers, and the normal condition of the crystal is disturbed.

Figure 4 c and d shows crystals with runs. The surface has melted because the temperature has risen sharply, or because the charge feed rate has dropped. The run sometimes extends right round the crystal (Fig. 4c), though in most cases the run is at one side only (Fig. 4d). In either case, the liquid soon runs away, and the runs almost

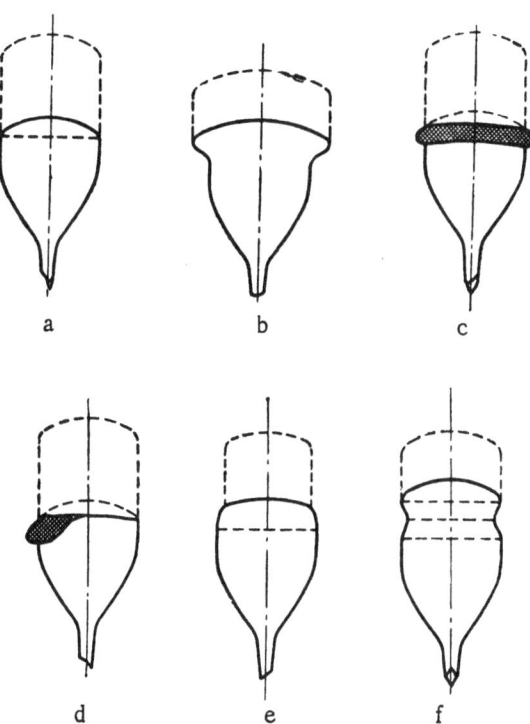

a   b   c

d   e   f

Fig. 4. a) Correctly grown and b) expanding crystals; c, d) crystals with runs; e) contracting crystal; f) crystal with a neck.

at once enter a region whose temperature is below the melting point of corundum. The result is to produce a suddenly cooled polycrystalline mass.

The film of melt left behind is much thinner than usual. The result is that a surface which has already been hot once is again heated to a high temperature. The effect of this is to produce the same flows as are found when a crystal expands.

The crystal grows narrower if the heat supply is reduced, because the edge cools (Fig. 4e). The heat supply may fall if the charge feed-rate (or the withdrawal rate) is increased. Here the charge forms a thicker layer over a lesser area, the linear growth rate rises, and the time any part is in the hottest zone is decreased. The amount of chromium taken up by the crystal increases, and the crystal becomes darker in color. Gas bubbles and unmelted material, as well as excessive stresses, may occur. The crystal shows a neck (Fig. 4f), if normal conditions are restored.

The Charge

The charge affects the course of the crystallization and the homogeneity of the crystal very much. Purity is important; so is uniformity in grain size. We have found that mixes made from "chemically pure" alum that has not been recrystallized are quite suitable. The charge is of better quality if the alum is fired in a muffle furnace instead of in an open-flame one. The grain size is smaller and more uniform.

Rod Corundum

At least 75-80% of the material in a boule is wasted as scrap in the cutting shop in the operations of slicing, cutting into blocks, and rounding to cylindrical blanks; cut ends and rims form the scrap (Fig. 5). These operations consume up to 40% of the diamonds used by the whole process. They are also very laborious, and result in a large holdup of material.

I considered how to eliminate this unproductive waste of corundum, diamond, labor and machine time, and set out to make the monocrystals as long thin rods of given size (Fig. 6).

The problem to be solved in making ruby rods are many and complex, because the conditions have to be controlled very much more carefully than is the case for boules.

Corundum rods of sizes between 1.5 and 3 mm can be made only if the temperature conditions and growth rate are controlled extremely accurately. An improved and semiautomatic apparatus was needed in order to make rods whose end surfaces were 50-60 times smaller than those of boules, and whose linear growth rates were 10-15 times larger.

The apparatus was designed and made at the Institute of Crystallography, Academy of Sciences, of the USSR. Electrolytic oxygen and hydrogen were used with this apparatus to make thin rods of good shape free from gas bubbles and other inclusions [1]. It was quite possible to produce these rods industrially by using the improved apparatus and pure electrolytic gases.

Pure gases are needed because the path from the jet to the end of the rod is 4-6 times smaller for rods than it is for boules. The conditions for preheating in the charge are thus much worse. The purest possible gases of the highest heat output are needed to make thin rods, because the charge must be heated and all gases eliminated within a very short length.

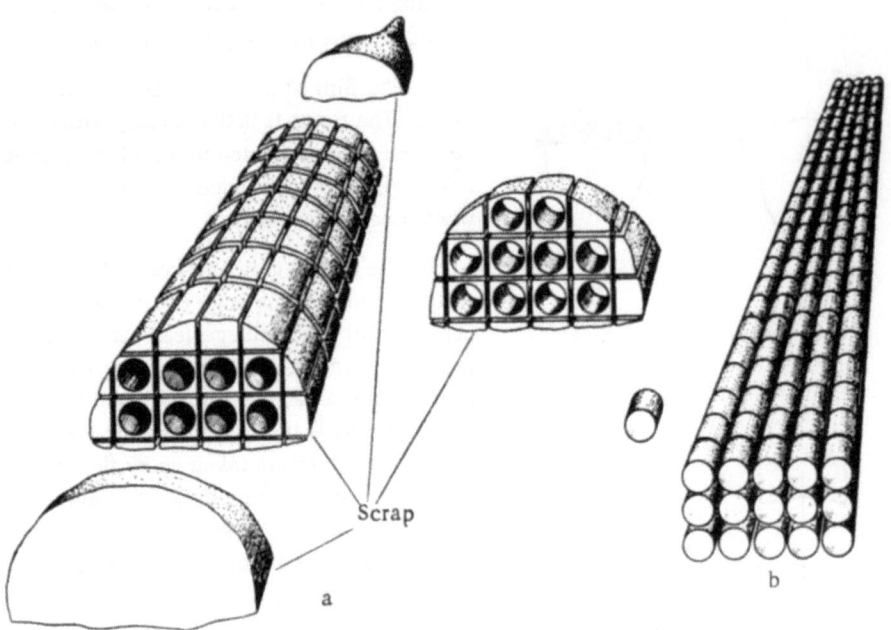

Fig. 5. Cut pieces from a) a boule and b) from rods.

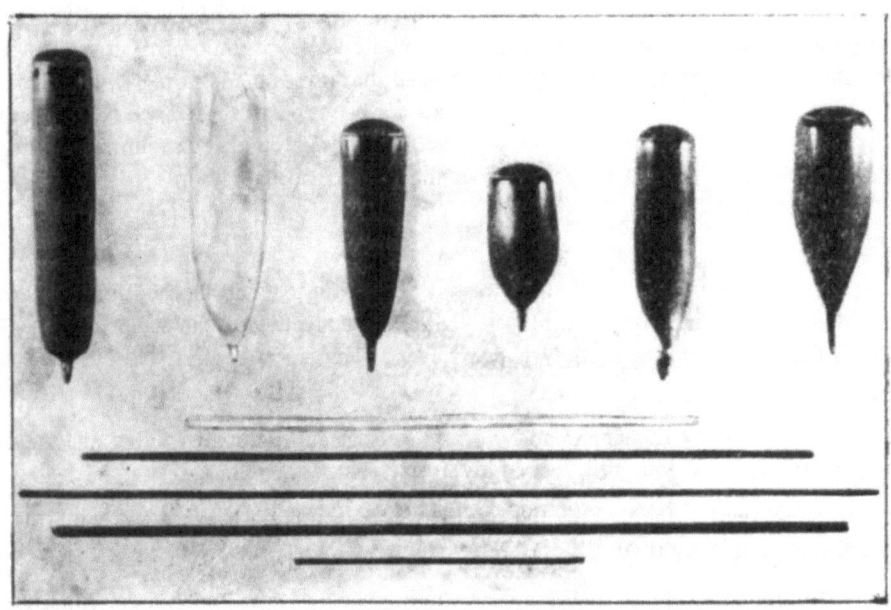

Fig. 6. Ruby and corundum as boules and rods.

Oriented Crystallization

The quality of a synthetic corundum part (ease of scratching and chipping) depends on the crystallographic orientation of the part.

If a boule consists of only one crystal that has arisen by chance, the relation between the crystallographic and growth axes is uncontrolled. Corundum has planes of imperfect cleavage on the rhombohedron faces, and the crystals crack most readily along such planes.

It is best to ensure in advance that the most favorable orientation is produced, in order to provide the greatest strength and to reduce the amount of scrap produced in making bearings, pivots, and so on. The most suitable orientation depends on the function of the part.

There are differences between the methods of growing oriented boules or rods and the methods used for ordinary boules grown from seeds. The seed is a thin round or square rod 1.5-3.0 mm in diameter and 20-30 mm long; it is held in a refractory holder coaxial with the burner jet. (The two axes must coincide as accurately as possible). This seed is used to cause oriented crystallization.

The seed is made by establishing the orientation of a rough-sawn boule by optical and x-ray methods, which boule is then marked and is sawn up into rods with a diamond saw. The rods are then of known orientation. This operation need be done only once, because future seeds may be cut from the rods grown with that orientation.

These rod seeds not only give oriented crystals; they also simplify the whole process, because no initial cone need be grown. Oriented crystallization on rod seeds has been adopted industrially. The recommended orientation is that in which the growth and optic axes lie at 57° to one another.

Improved Apparatus for Growing Corundum Monocrystals as Boules and Rods

The apparatus to be described provides the right conditions for correct crystallization and exact growth control.

The first industrial apparatus (developed in 1937-1939, Fig. 7) was meant to make boules only. In 1939, this apparatus was introduced industrially, and the first industrial synthetic corundum was made with it. In 1945-1947, an improved industrial plant was developed for growing corundum rods. This plant was also suitable for making high-grade ruby boules. In particular, it was much better for growing dark ruby boules, which are much more difficult to make than the light-colored ones.

The apparatus will be described. Some parts of the original plant (the SP-3) will also be described. This plant is still of value as a simple laboratory equipment. It is also suitable for industrial use in processes in which extremely accurate control is not needed.*

General Systems of the Support, the Drive and Crystallization Unit

Figure 8 shows the kinematic system of the new equipment.

The apparatus is designed to have a light frame, which is fixed at the level of the valves (about 1 m high), and is connected to the floor. The system is designed to work with 26 sets of apparatus (13 on each side) in a bank. On each of the banks there are two rotating transmission shafts and two rocking shafts. The entire apparatus is driven by a single motor and is noiseless in operation. If rods are to be made, the number of rocking shafts (4-8) must be such that the holder is withdrawn at 100 mm/hr when the speed regulator is set at the center position. The speed is 10 times lower for boules, assuming a growth rate of 40 carats per hr.

The drive system is such that the speeds of the shafts on the two sides can be adjusted independently. It is thus possible to grow boules on one side and rods on the other. The bearings, cams and followers are protected against dust and contaminants.

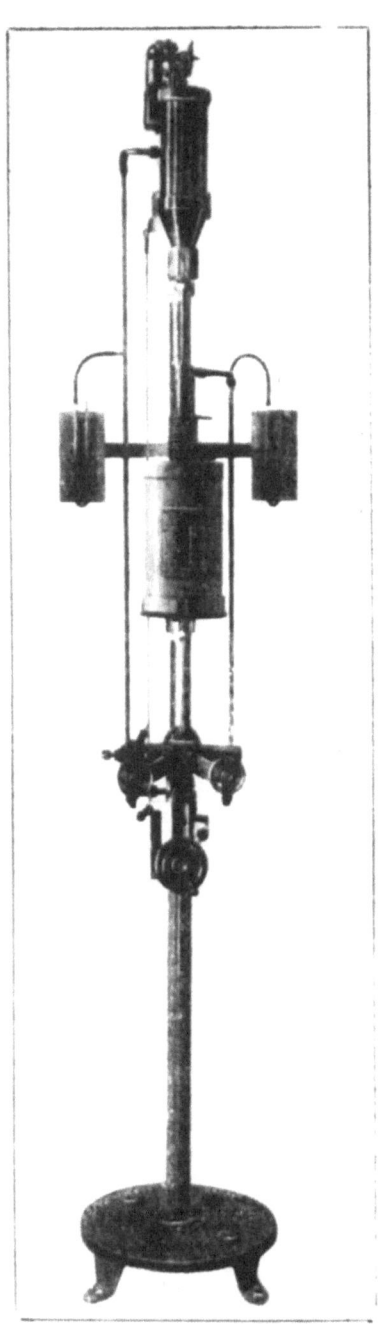

Fig. 7. The crystallization apparatus of 1939 for growing corundum boules.

* Popov developed four distinct parts of the apparatus: the SP-1 burner in 1933; the SP-2 crystallization unit in 1934 (both of these were meant for oxygen and town gas); the SP-3 industrial plant in 1938 (for boules, with oxyhydrogen burners); and in 1947 the SP-4, the automatic equipment for boules and rods – Compiler.

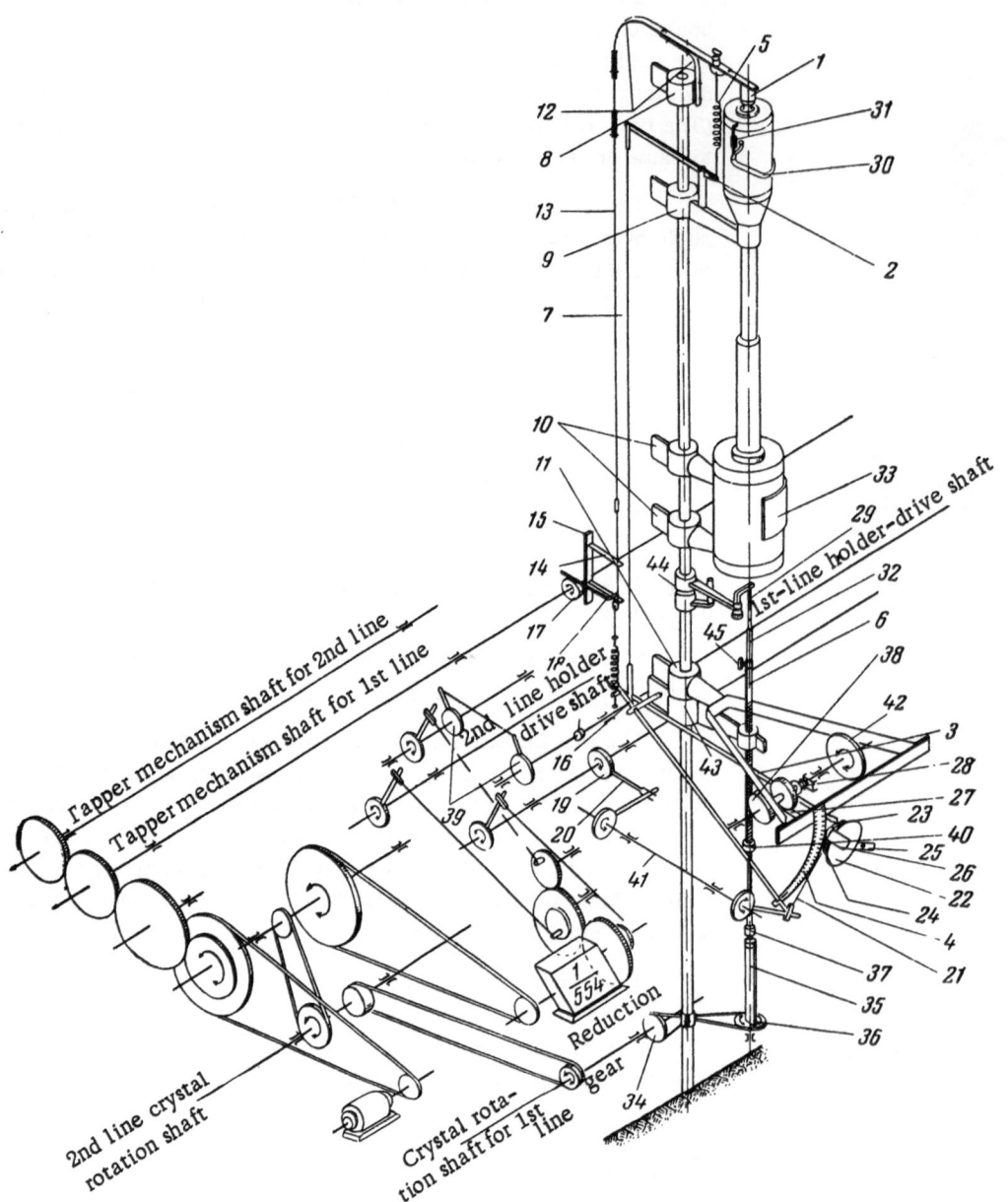

Fig. 8. The kinematic system of the apparatus for growing ruby and corundum: 1) tapper;
2, 3, 16, 30) levers; 4) scale; 5) spiral springs; 6) lead screw; 7, 15) shafts; 8) boss;
9) boss for head; 10) boss for oven; 11) boss for valve system; 12, 14) flat springs; 13,
21) link wires; 17, 19, 39) shafts; 18) rocking follower; 20) fork-and-pin mechanism;
22) ratchet mechanism; 23) dog; 24) lid; 25) handle; 26) axle; 27) worm; 28) crown
wheel; 29) crystal; 31) tensioner; 32) "candle"; 33) slide; 34, 36) pulleys; 35) steel
scale; 37) end-stop; 38) gear; 40) extra load; 42) small handwheel; 43) boss for holder;
44) viewer; 45) clamp.

The main feature of this bank is that each unit is fixed separately. There is thus no transmission of vibration from one unit to another, which prevents any disturbance to the crystallization.

Figures 9, 10, and 11 show the improved apparatus for growing corundum boules and rods. The main units are the tapper mechanism, the head and bunker, the charge feed device, the burner, the crystallization furnace (with a device for closing the viewing port as the charge feed is cut off), the automatic holder, the holder drive, the valve gear and the automatic expander.*

---

* The units are described in sequence from top to bottom.

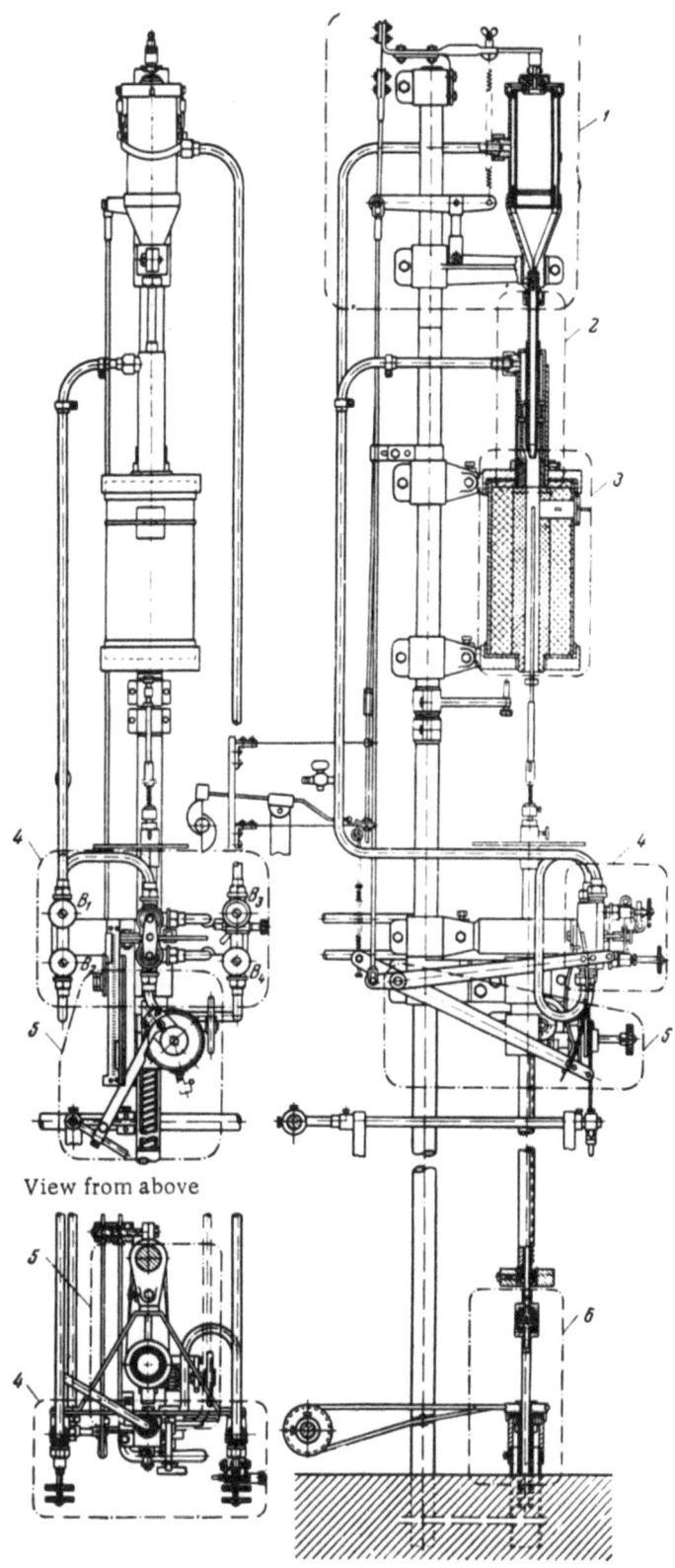

View from above

Fig. 9. The improved crystallization apparatus: 1) head (see Fig. 14); 2) burner (see Fig. 9); 3) oven (see Fig. 19); 4) valve gear (see Figs. 26 and 28); 5) holder (see Figs. 22 and 23); 6) drive for turning holder (see Fig. 24). B_1 and B_2 are hydrogen valves; B_3 and B_4 are oxygen valves.

Fig. 10. The improved apparatus for growing ruby rods and boules.

The Charge Feed System and the Shaker Box

The system for controlling the feed consists of several separate parts, namely the tapper, the shaker box, and the control box.

The tapper. The accuracy of the feed is related to the growth rate, and so the tapper must not be affected by backlash, by wear caused by the very abrasive alumina, and by lack of adjustment in the moving parts.

The tapper has a simple function. It taps the shaker box continually with a precisely adjustable force, in order to cause the charge to pass into the apparatus at the right rate. It is very important to keep the force constant. The flow of charge varies if the tapper does not work stably, and the crystals show bubbles and other flaws.

Several types of tapper have been used, namely electrical, electromechanical and purely mechanical. Each type has its advantages and disadvantages, all of which ultimately affect the grade of the crystal.

Mechanical tappers are at present the most reliable for routine use.

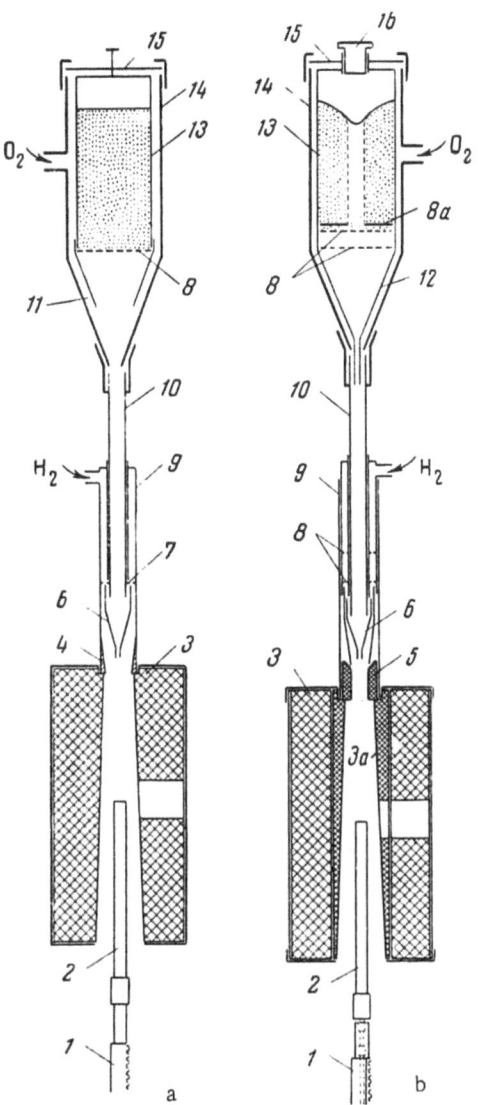

Fig. 12. The electromechanical tapper: 1) arm, 2, 3, 7) springs, 4) cam, 5) damper, 6) roller, 8) plate, 9) boss, 10) wingnut, 11) ring.

Fig. 11. Comparison of apparatus: a) old-style apparatus, b) Popov's system; 1) holder rack, 2) holder, 3) oven, 3a) internal liner, 4) metal nozzle, 5) ceramic nozzle, 6) tip, 7) mesh, 8) grid, 8a) diaphragm, 9) burner, 10) oxygen tube, 11) guide-cone, 12) cone and guide-tube, 13) internal vessel, 14) head, 15) membrane, 16) filling hole.

The usual types of tapper are briefly as follows.

1. An electric tapper driven by a central common switching unit is simple, compact and readily repaired; each unit has to be fitted wiht its own feed-rate control.

Tappers with their own switching are more complex and less convenient, and have not been found to be very reliable.

Both types are unstable, in that the blows vary in force, and so the feed is unsteady. Voltage fluctuations also affect the accuracy.

In the SP-3 we used a good electromechanical tapper (Fig. 12), whose mechanism was controlled from the control panel. The main part was the small electric motor, about which all the other parts were built up. The tapper mechanism was as follows. The cam with four pins 4 is fixed to the motor spindle. The tapper arm 1 is fixed to the plate 8, which has a rubber buffer 5 (to reduce the noise). The roller 6 is fixed to the stand, and the pins slide over it. The lever 1 is held by the spring 7. The force of the blow is controlled by the spring 3, and by the return spring 2. The tapper is mounted on the boss 9, which is fixed by the wingnut 10. The wingnut is slackened off and the tapper is moved to one side when the charge is being inserted. The stop ring 11 ensures that the tapper returns exactly to the initial position.

2. Mechanical tappers are simpler than electrical and electromechanical ones. They can be worked stably and reliably from a drive. They are simple to make and repair.

In industry such tappers are often made badly and work unreliably, however. The drive system tends to be cumbersome and to jar the apparatus. This jarring hinders the crystallization.

To remove these faults we designed a new, simple and compact tapper (Figs. 9 and 10). This design differs from all others in that certain important rotating and pivoted parts, which rapidly and inevitably wear, have been replaced by rocking parts.

The design is such that the feed is very accurate, and the machine as a whole works reliably for long periods before repairs are necessary.

The tapper mechanism (Fig. 8) must have a tapper that always works accurately. In our design the tapper 1 is fixed to a flat spring 12 held in a boss 8, and is coupled by a link wire 13 and spring 14 to the shaft 15, which is fixed to the common framework. Springs 12 and 14 are adjustable. The tapper can be balanced exactly, so the least tension in the spring 5 causes it to strike the shaker box regularly and accurately. This device is worked from the rocking follower 18 and spring 14, which has a rubber damper. The tapper is fixed to the springs along their planes, and so comes to balance (floating action). Springs 12 and 14 are not deformed, because the working load is so small, and so the tapper works accurately and stably over long periods. The tapper can be demounted in order to recharge the feed vessel without disturbing the feed rate.

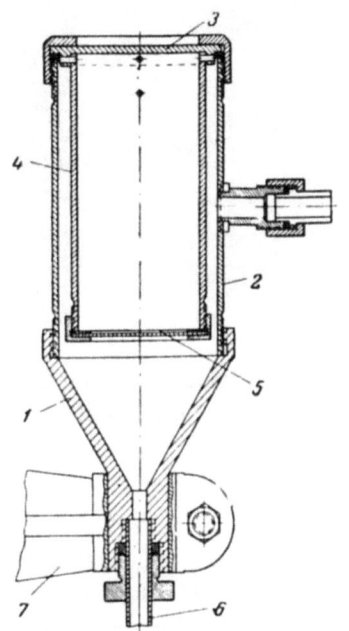

Fig. 13. The head of the SP-2 apparatus:
1) cone, 2) outside cylinder, 3) base,
4) shaker box, 5) grid, 6) tube, 7) boss.

The head and shaker box are designed to feed the charge to the flame. The quality of the crystal depends very much on how well they work. We describe the working of the shaker box on the original laboratory design (SP-2, 1933, Fig. 13)[*] in order to show up the main design features.

The box 4 is filled with charge, and is fixed rigidly and hermetically to the outside cylinder 2. The taps on the top 3 cause the powder to shake out through the grid 5 onto the cone 1; it then enters the tube 6. The charge must enter in strictly controlled amounts, and must not be held up at any point. In this system, however, it often settled and became held up, and then rushed down. These rushes of charge were not melted by the burner, and so the crystals had unmelted spots and bubbles.

The charge settles and sticks at the joint between the cone 1 and oxygen tube 6, because the surfaces are slightly uneven. The powder is also held up in the cone 1, which is held fast in the boss 7, and does not vibrate adequately even if the tapper strikes hard. The surface of the cone is difficult to polish to an adequate finish. Even a well polished cone soon becomes scored by the continuously falling and sliding charge. The cone has to be taken out from time to time and polished — a very troublesome operation. The centering on reassembly is lengthy and complicated. The charge coming from box 4 should not contain large clumps of particles. The grid 5 has to be very fine (1600-2500 holes per cm^2). The box is usually held fast in the head. The taps then have to be so hard that the whole apparatus tends to shake. Other equipments that are not working may even be seen to release some charge. The taps also shake the burner through which the oxygen and charge are passing. The burner vibrates both horizontally and vertically, and so the charge is thrown about; charge is wasted and the crystal becomes wider.

Several types of shaker box were made during the laboratory studies. All the above defects were removed in the final design. This design provided a completely uniform flow of charge to the rod. The design was as follows.

The vessel 1 (Fig. 14) was suspended from the head by the thin membrane 2, which design ensured that light taps from the tapper 3 produced a marked vibration, adequate to cause the charge to pass through the grid 5. The

---

[*] This type was also later used in production work.

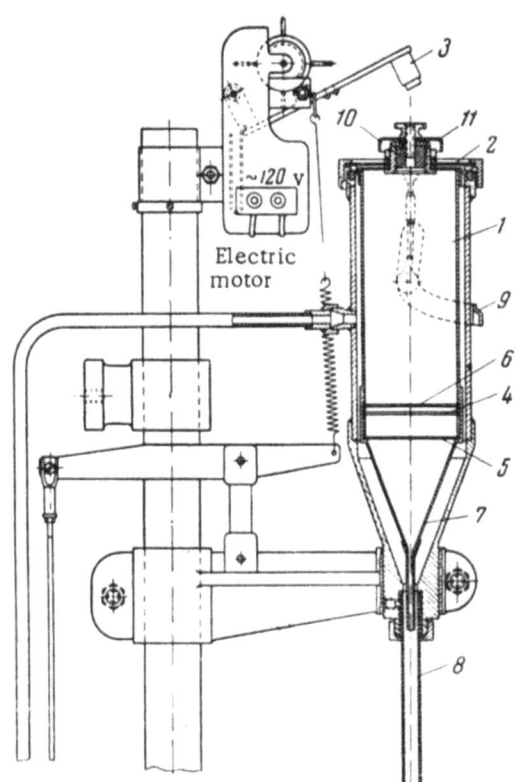

Fig. 14. Head with shaker box and individual tapper mechanism: 1) vessel, 2) membrane, 3) tapper, 4 and 5) grids, 6) diaphragm, 7) cone, 8) oxygen tube, 9) lever, 10) seal, 11) protective cover.

charge was made to flow quite evenly by the diaphragm 6 (which had small holes in it) and the grids 4 and 5, placed one above the other. The diaphragm held back the packed charge. The tapper caused the loose top charge to pass through the holes onto the first grid. The charge did not become packed in this section. The flow of charge was controlled by the size of the holes. From the first grid the charge flowed almost uniformly to the second grid, and about one-third filled the space between the two grids. The charge above the second grid was always thoroughly loosened, and so flowed through quite evenly.

The shaker box was fitted toward the bottom with a light cone and tube 7 which entered the oxygen tube 8. This prevented the charge from sticking. The cone and tube shook together with the box, and so any charge present could not become stuck.

The tube has to be set exactly in the center of the lower hole. The job of reassembling the head at a later date is simplified by marking the head, membrane and box at appropriate points. The charge must be prevented from getting onto the outside of the box during assembly. The apparatus will then work properly and will not become blocked.

A charging hole with a rubber seal 10 (Fig. 14) is fitted. Metal lids had been used in the earlier design; these tended to contaminate the charge with metal when they were screwed on.

The special cover 11 prevented the rubber seal from becoming contaminated. The charge entered the shaker from a special supply vessel (Fig. 15) via this seal. This vessel ended in a funnel, which entered the box to a set level.

This design of seal and supply system prevented the charge from being contaminated. The vibration system shown in Fig. 15 was used to fill the shaker, and to prevent the charge from being contaminated in the crystallization shop. A lever-operated seal was also fitted (Figs. 12 and 14); a single movement of lever 30 opened or closed the vessel. The pressure applied to the seal was controlled by the tensioners 31 (Fig. 8), which had right-hand threads only, and by the thickness of the rubber packing.

Charge feed-control. The feed-control system did not provide the required accuracy. The spring tensioning mechanism on the tapper worked unreliably, did not give accurate control of the feed, and had to be removed with the tapper during loading, which disturbed the settings. The control mechanism was also mounted on or near the tapper. This position is inconvenient, because observations on the crystallization had to be stopped to do this adjustment, and the feed had to be adjusted with neither the charge nor the crystal visible. The control mechanism had no scale to assist in making settings, or in resetting after dismantling.

In our apparatus, the feed was controlled as follows. A lever system 2 (operated from the control panel) was used to make fine adjustments. The spring 5 couples this level to the tapper 1. Lever 3 is used to make coarse adjustments, which lever operates on 1 and 5 via the link 13. This system has a scale 4, which is used to read and set the flow accurately. The charge can be seen while adjustments are made, and the quality can be judged from the flow rate.

This centralized adjustment in accordance with an exact scale simplified the work and improved the crystals.

The Burner

The burner is the main part of the crystallizer. A high flame temperature (over 2000°C) is needed. The flame must also be large horizontally and vertically if large monocrystals are to be grown. The pressure drop in the working area of the flame must be very low if the gas pressures are low (40-50 mm of water, at the burner jet).

The following design points are important.

1. The construction must be simple and accurate.

2. The gases must mix as completely as possible in the burner. This ensures that the flame temperature is uniform.

3. The gas must burn without disturbance.

4. The charge must not get held up in the joints in the oxygen tube, or on the walls in the head.

5. Repairs and replacements must be effected easily and simply.

6. Recentering should not be necessary when the burner has been repaired or adjusted.

7. The burner must operate for long periods without needing repair.

There are several designs of oxyhydrogen burner. They differ slightly in design and dimensions.

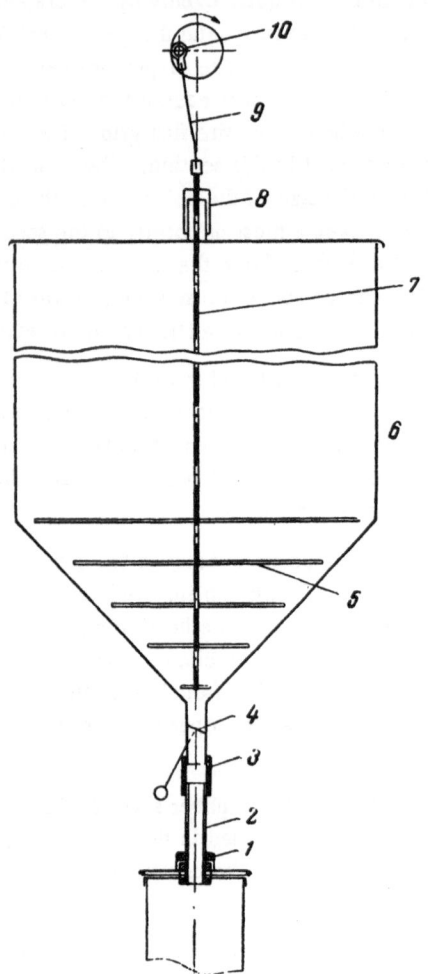

Fig. 15. Charge feed device: 1) rubber seal, 2) glass tube, 3) rubber ring, 4) valve, 5) duralumin grid, 6) bunker, 7) rod, 8) filter, 9) spring, 10) eccentric.

A general fault of many such burners is that the gases do not mix properly. Burners with large metal nozzles (Fig. 11) and no mixing chambers have two distinct streams of gas, because the speeds of the two gases are very different (by a factor of 15-20). The mixing is not good. The length of the central cone of cold oxygen in the flame is increased. The charge has farther to travel in the cool zone of the flame. The charge does not melt completely in the flame, and so the crystal shows many faults. The gas consumption is excessive, and the joint between the oxygen tube and the head tends to hold back the charge. In some designs the jet is screwed into the oxygen tube, while the oxygen tube is connected to the head via a conical joint. This gives rise to two sets of obstacles, and to leaky joints.

The burners do not last very long because the oxygen jets and nozzles scale up. The frequent recentering required on replacing or repairing the burner consumes much time.

The burner described below (Fig. 16) was designed in the light of the design points listed above.

The main cylinder 2 has an endplate 3, into which is soldered a guide tube 4 and a side arm 7, through which the hydrogen comes in . The space formed by the cylinder and tubes is fitted with a grid 14. This grid smoothes out fluctuations in the gas pressure, prevents blow-back, and ensures that the gas enters the mixing and combustion chambers evenly. The chamber 2 is held in the outer sleeve 1, which holds the nozzle 15.

The oxygen tube 5 slides freely in tube 4; tube 5 has a cone to the head 19, which is held by a cylindrical joint and pressure ring. The charge cannot be held up, and the head is sealed. The centering is not lost when the burner is dismantled. The bottom of the tube is threaded to take the jet 6. This jet is conical (as well as internally threaded) and passes smoothly into the cylindrical section (which is accurately gauged). Tube 5 and jet 6 are so joined that no charge can be held up.

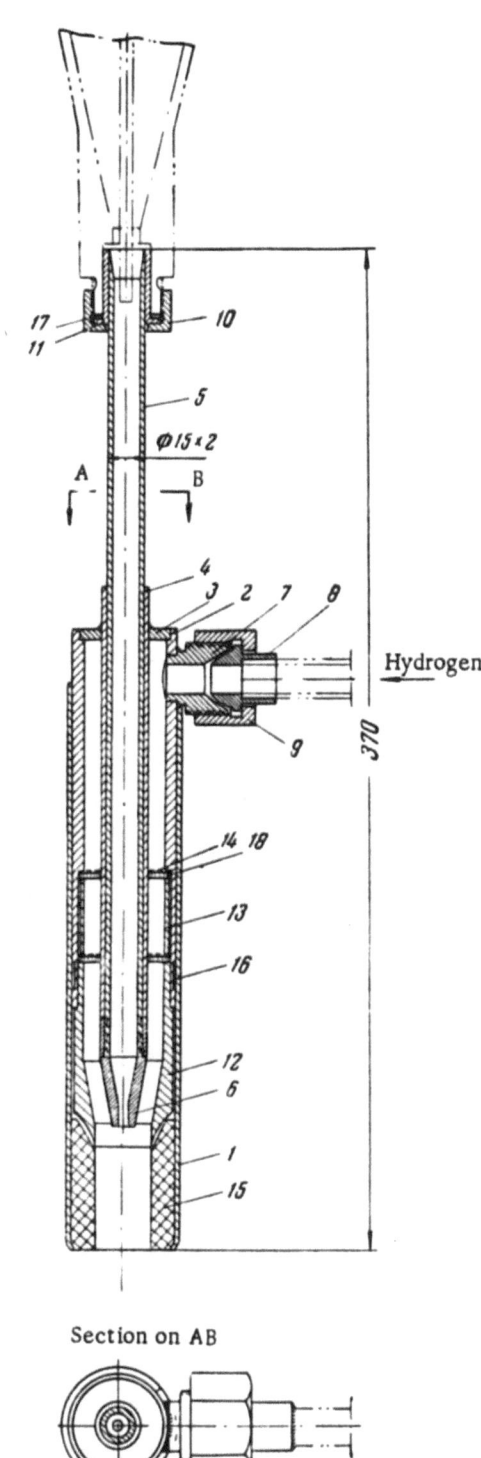

In this design the mixing and combustion chamber (nozzle 15) is made of a refractory ceramic (corundum, corundized alumina, etc.). The ceramic is worked fairly easily. The shape of the nozzle is chosen to aid mixing. The ceramic protects the burner from corrosion and rapid wear.

Scale usually forms on the oxygen jet when explosive mixtures are burned. Charge tends to stick to this scale. The direction of the stream of oxygen is affected, and the charge swirls about as it falls. The crystal is distorted, and the consumption of charge is increased. This fault is absent from our design, because the jet was made of heat-resisting steel (ÉI-292 or Kh-25-Yu-5); no charge became stuck. These jets were stable, required fewer shutdowns, and ensured that the gases flowed smoothly.

The heating power and mode of use could be varied widely without redesigning the burner. The size of the mixing and combustion chamber could be adjusted by moving cylinder 1 relative to the body 2; the sizes of the refractory nozzle 15 and of the jet 6 could also be varied.

The burner of Fig. 16 was worked as follows. The hydrogen entered the upper part of the body 4 via the side arm 7. It passed through the grid 14, and entered the second part at a lower pressure. The second grid had a certain resistance, and so the gas entered the mixing and combustion chamber smoothly at a pressure excess of only 2-4 mm H_2O. The hydrogen met the oxygen entering through jet 6 at a pressure excess of 40-50 mm H_2O. The oxygen reached the jet through the head and the oxygen tube 5. The burning gas passed from the mixing chamber into the crystallization furnace.

The Furnace

The design of the furnace determines whether the working space is evenly heated, whether the charge falls smoothly without swirling, whether the gas burns evenly, and whether the heat is properly retained in the crystallization space.

There are two designs of furnace for corundum. They are both very simple.

The first type consists of two firebrick halves (Fig. 17), which, assembled, form a thick-walled hollow cylinder of outside diameter 100-120 mm, of inside diameter 40-60 mm, and of height 150-250 mm. The shape is sometimes square. The second type is also made of firebrick, but the cylinder is in one piece (Fig. 18).

In the first type the burning gases pass through the cracks between the parts and through the asbestos insert used as a lid, especially through the gap between this insert and the burner. Much heat is thereby lost, and the temperature in the furnace becomes uneven.

Fig. 16. The burner: 1, 2, 4 and 5) tubes; 3) end-plate; 6) jet; 7) inlet side-arm; 8) nipple; 9 and 10) pressure rings; 11 and 15) inserts; 12) refractory insert; 13) ring; 14) grid; 16) asbestos filler; 17) packing; 18) ring; 19) head cone.

Section on AB

Fig. 17. Demountable halves of a firebrick furnace.

The brick is damaged if the halves are moved every time the burner or crystal is examined; this prevents the furnace from being set up accurately.

The material used for such furnaces is inadequately resistant to the heat. The parts warp, crack and fuse internally, and finally fall to pieces. The cracks and runs make the internal surface rough, which sets up turbulence; the gas and charge are used wastefully, and the quality of the crystal is affected. These demountable furnaces soon fail.

It is almost impossible to center a demountable furnace exactly, because it is very laborious to set each half to be strictly vertical every time. Good centering is, however, essential to making good crystals. The furnace must be set in the proper position relative to burner and holder to ensure even heating and the correct mode of supply of charge to the crystal.

The second type of furnace is better than the first. The heat losses are much smaller, and the centering is

Fig. 18. Furnace for growing corundum boules.

much easier. However, to check the burner, and to set up the furnace, either the burner or the furnace must be withdrawn. The entire system must then be recentered, which makes the operation tedious. Integral furnaces crack and run internally in the same way as demountable ones. They are stronger and last longer, however.

We now deal briefly with the furnace support. The apparatus will not work well, even if the furnace is good, if the support is not firm. If the support tilts through bad design or manufacture, the furnace tilts also, and the centering is disturbed. The support must work accurately and stably, and must not distort at high temperatures.

The furnace must have the following features:

1) it must be easy to take down and repair;

2) it must waste little heat;

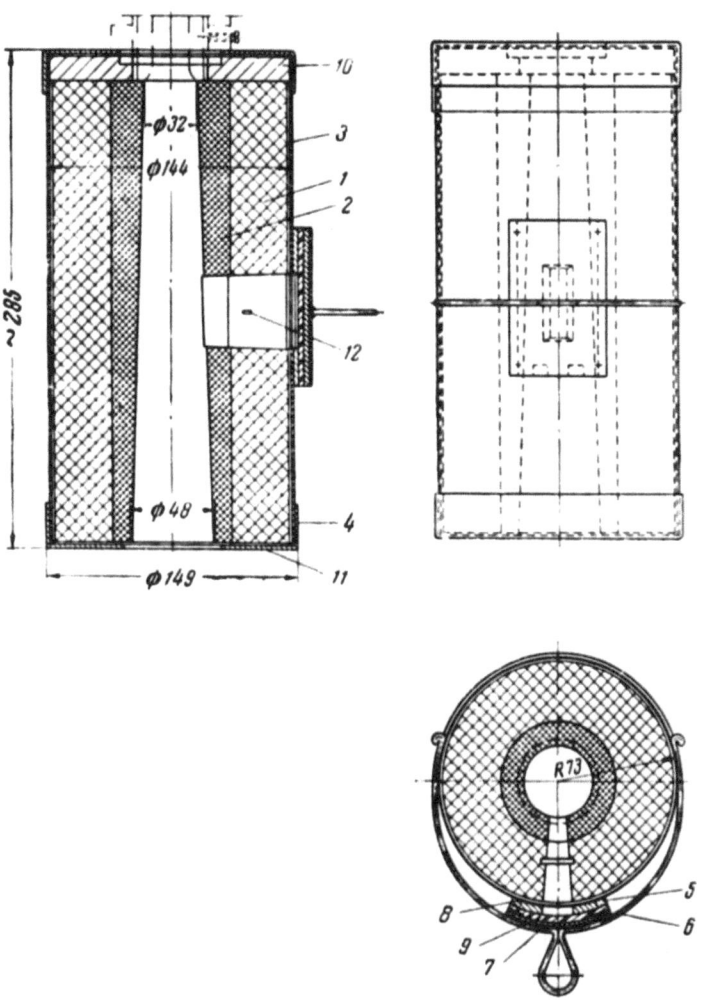

Fig. 19. Section of a furnace for growing corundum boules: 1) outer brick, 2) inner
lining, 3) case, 4, 10) rings, 5) insert, 6) holder, 7)slide, 8) rivet, 9) packing,
11) endplate, 12) viewer.

3) it must consist of parts that are easily and rapidly replaced;

4) it must be easy to center accurately to the burner and holder;

5) it must have a long life.

We have developed a furnace that satisfies these requirements (Fig. 19). A primary need was for a good refractory material, which would be stable and of low thermal conductivity.

The design was more complicated than in the cases above. The cylinder 1 (Fig. 19) was made of an insula-
tor and was in one piece. The liner 2 was of appropriate shape and was made of corundum. The ceramic parts
were held in a thin stainless steel jacket.

An observation port was fitted, as in other designs. The viewer 12, of heat-resistant material, was used to
view the focus of crystallization when growing rods. For use with boules the viewer was a three-point sighting
device (see Fig. 3). The viewer was fixed opposite the viewing port and was supported by the lower furnace boss.

The port was closed by a stainless steel slide 7 fitted with an asbestos lining. The slide was set in a special
device (not shown in Fig. 19) coupled to the charge feed, and was closed before the apparatus was shut down. The
parts of this device were fixed in part to the lower furnace boss and in part to the common framework. The port
was closed by a lever worked by a flat spring. The lever 3 (Fig. 8) was moved up over the scale to the stop at the
same time as the feed was cut off.

This type of furnace heated up quickly and evenly, and cooled slowly, which provided good conditions for the crystal to cool in.

Only the inner lining need be replaced when the furnace breaks down. This lining can be replaced several times.

Demountable bosses (Fig. 8) were used to hold the furnace, which were of great assistance in rapidly and exactly centering the furnace and burner.

The Automatic and Manual Crystal-Holder Drives

The drive is used to insert the crystal holder ("candle") in the furnace at the start, to lower the crystal gradually during growth, and to withdraw the finished crystal at the end. The drive must work exactly and smoothly during the growth, and must be adjustable for speed. Runs and distortions occur in the crystals if the drive does not work properly.

The feed and heating conditions must be uniform if good crystals are to result. The crystal must lie exactly in the center of the furnace in order to ensure this. The temperature gradient will otherwise not be the same at all points on the growing surface. The candle is held in a holder, which is fitted with means of centering the crystal exactly. This holder must be fixed rigidly to the drive system. Any play in the holder makes the apparatus unusable.

The centering must be simple to set and must be accurate; it must remain so, as the holder is moved vertically.

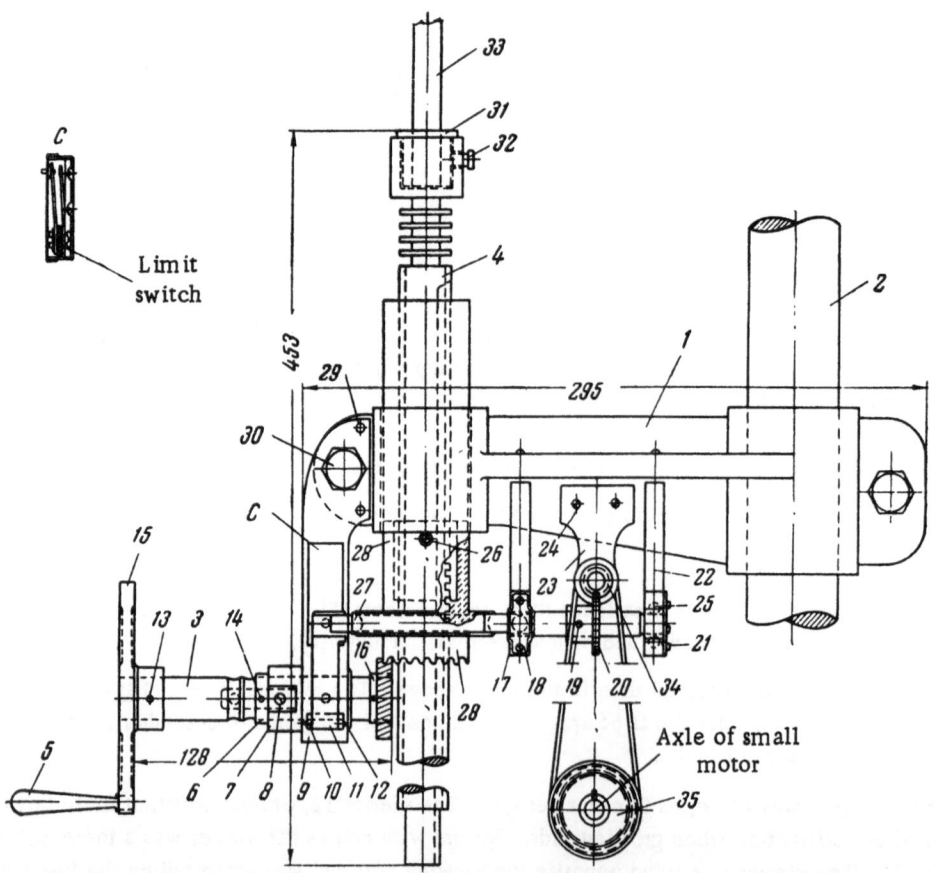

Fig. 20. The automatic drive system: 1, 9, 17, 22, 23) support arms; 2) rod; 3) shaft; 5) handle; 6) ring; 7, 31) bushes; 8) spring; 10) loop; 11) lever; 12, 13, 29) pins; 15) handwheel; 16, 20) gears; 18) cover; 14, 19, 21, 24, 26, 32) screws; 25) ball bearing; 27) worm shaft; 28) compound bush; 30) bolt; 33) candle; 34) worm; 35) friction adjuster.

The drive system must be protected from dust by enclosing it in a cover made of thin noncorrosive metal, as far as is possible.

The main faults in drive systems are as follows.

1. The control is manual instead of automatic. Manual control means that the crystal is lowered in steps at arbitrary intervals. The crystal is affected unfavorably by this. Runs, rims and bands result.

2. The drive is mounted on a separate stand or base, instead of on the framework common to the rest of the apparatus. This makes repair and assembly more difficult, and exact centering almost impossible. The holder does not stay in the center as it is moved up and down.

3. The refractory candle is held in a holder which can move freely on the sleeve fixed to the drive shaft. This violates one of the basic conditions for proper crystallization.

Two forms of automatic drive were developed for the industrial plant in order to eliminate these faults.

In the first (Figs. 20 and 21) the automatic drive is provided by a worm system fixed to the drive shaft, which worm is coupled to a reducing gear. The latter is coupled to a frictional speed adjuster, and the crystal is lowered smoothly. This drive works as follows (Fig. 20). The power is supplied from a common shaft, or from an individual electric motor, via the friction adjuster 35, and thence via the worm 34 to the worm gear on the compound bush 28, which turns and drives the lead screw 4, and thereby moves the crystal. The rate at which the crystal is lowered is adjusted by varying the speed. The friction adjuster may be omitted; the system then works at a single speed.

In the second system (Fig. 22) the crystal was lowered in steps at an adjustable rate by using a rachet and pawl system (Fig. 23), with an adjustable pawl. Here again the drive was provided by an individual electric motor, or by the transmission system, in which the rocking shaft system was used (Fig. 8). The rocking motion required by the link wire 21 and dog 23 was transmitted via a gear train from an individual motor mounted on the apparatus. In our case the rocking motion was transmitted from the rocking shaft 19 via the bush 20 and geared shaft 41. The dog 23 slides in the cover 24 and engages the exposed teeth on the gear 25; it thus drives the axle 26 and the fine-feed manual handle. The worm 27 is force-fitted on the axle 26 and transmits the motion via the worm gear 28 to the spur gear 38 (concealed in the mechanism). The worm gear and spur gear are fixed on the same shaft as the handwheel 42, which is used to move the crystal holder rapidly. The spur gear drives the lead screw 6.

An oblique tooth and load 40 are fitted to the lower end of the lead screw, in order to prevent sticking or uneven motion in the holder, which would spoil the crystal. The rate of movement is adjusted by moving the handle 25 in the cover 24 to uncover the required number of teeth on the pawl 22 (each tooth has a definite time-equivalent). The drive is stopped by covering the teeth completely. This device is readily adapted to any design of drive.

The holder is set in a vertical direction by using the adjustable clamp, which can be set very exactly. Covers (not shown) protect the mechanisms from dust.

The second design is simpler, more compact and more accurate, and complies with the very difficult requirements imposed by rods.

This design has the following distinctive features.

1. The drive shaft is triangular and has cross-cut grooves, which slide without backlash on inserts.

2. The shaft is designed to turn the crystal smoothly. This is done to even out the temperature and growth con-

Fig. 21. The automatic continuous-action crystal drive.

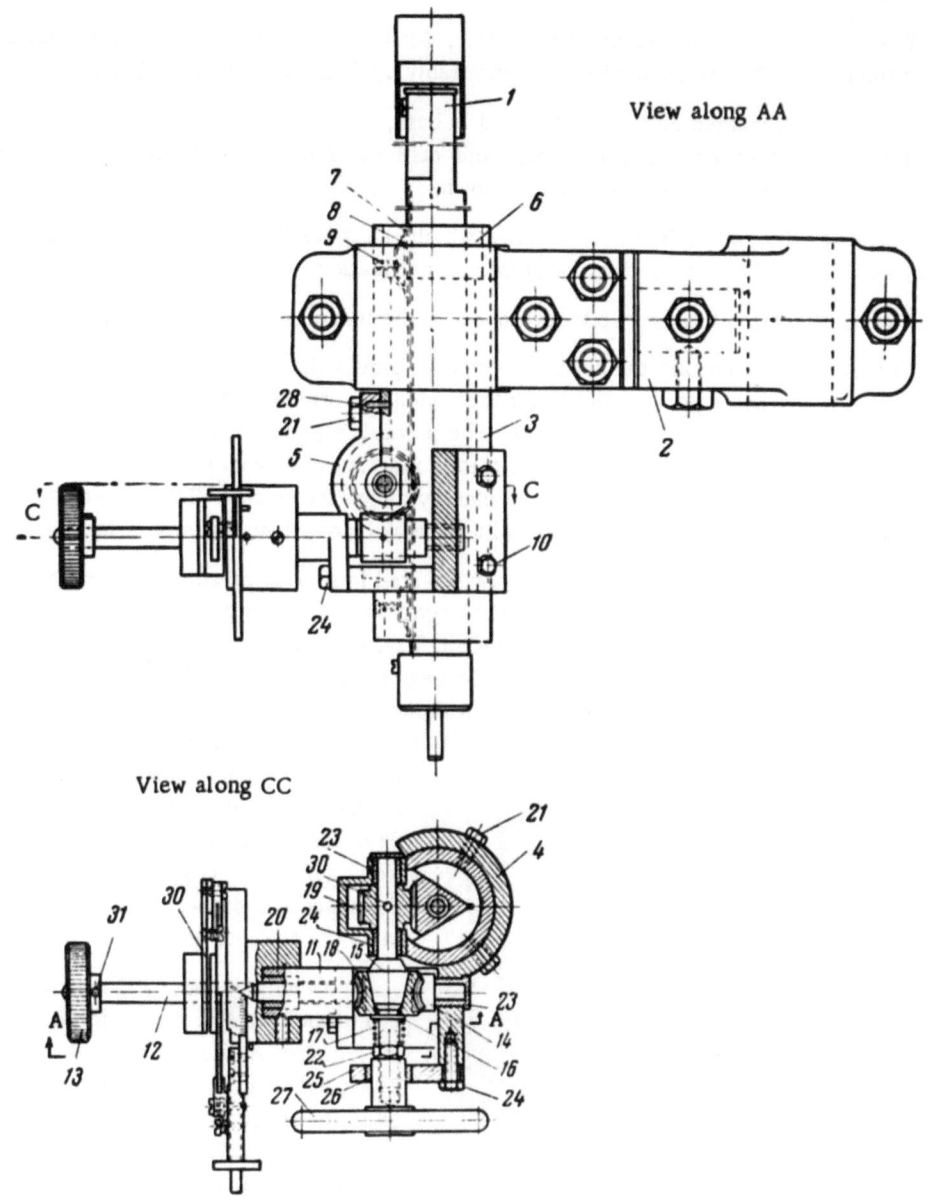

View along AA

View along CC

Fig. 22. The automatic step-action drive system: 1) shaft of holder; 2, 11, 25) support arms; 3) bush; 4) bush clamp; 5) cover; 6) guide; 7) slide; 8, 17) springs; 9, 10, 24, 31) screws; 12, 15) axle; 13, 27) handwheels; 14) worm; 16) pressure plate; 18) worm gear; 19) spur gear; 20) bush; 21) bolt; 22) nut; 23, 26) bushes; 28, 29) dowels; 30) drive unit.

ditions at the surface. The rods then grow as true cylinders which deviate from circular by 0.03-0.06 mm.

3. The drive that turns the crystal (Figs. 24 and 8) transmits the motion from an individual electric motor, or from the transmission, to the moving part of the shaft. This part consists of the steel strip 35, which is an extension of the axle that turns within the shaft 6. The strip 35 enters the slotted rod 36, within which it moves freely; it allows the shaft to work without jerks or vibration. The slotted rod 36, with the ball bearing and bush (crystal-holder drive), is mounted in a tube which is solidly fastened to the base and enters the ground.

124

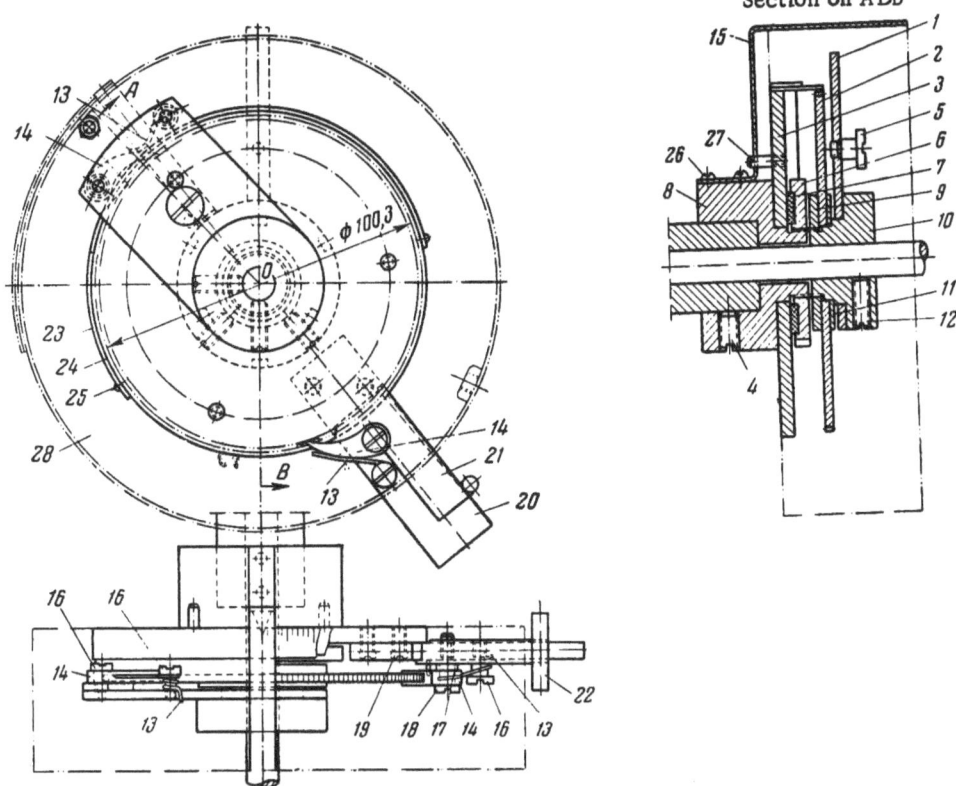

Fig. 23. The automatic step-action drive system: 1, 20) strips; 2) toothed ring; 3) disc; 4, 5, 12, 16, 18, 19, 25, 26) screws; 6, 9) nuts; 7) insert; 8, 10) bushes; 11, 17) spacers; 13) spring; 14) dog; 15) pointer; 21) lock; 22) rod; 23) plate; 24) scale; 27) pin; 28) cover.

The rotation of any one crystal can be coupled and uncoupled with the special clutch 37, which connects the axle within the shaft to the steel strip.

An improved system was used on the experimental equipment. The crystal was lowered manually until the automatic drive was introduced (Fig. 25).

Lately, several automatic crystal-holder drives have been introduced. In these, the drive is transmitted by friction. There can be large errors in speed, etc., such as jerkiness, variable speed, sticking, and so on. The speed cannot be set and controlled accurately.

When the crystal has been expanded up to size and has begun to grow as a cylinder, the surface has to be kept at the same point, i. e., at the point at which the expansion stopped. The growth rate must equal the rate of withdrawal. If the growth rate is known for given supply conditions, the withdrawal rate can easily be set appropriately.

A good manually operated drive may be of interest in certain contexts. The manual SP-3 system (Fig. 26) has proved useful in industrial syntheses. The wheel 4 is fixed to the axle 1. The other end of the axle is fitted with the cone gear 12, which engages with the compound bush 13 when manual control is used. This bush is coupled to the guide bush 3, and turns freely within it. The thread within the bush is in constant engagement with the lead screw 2. The top 5 is finned, which assists the air cooling of the lead screw and holder 7. The grub screw 8 holds the refractory candle 6 and ceramic skirt 9, which latter protects the lead screw from the heat and dust. The lead screw is prevented from turning with the compound bush 13 by the retaining groove 10 and retaining screw 11. In manual working the wheel 4 drives the bush 13 via the cone gear and moves the candle smoothly, without jerks upwards or downwards, while the centering is retained perfectly.

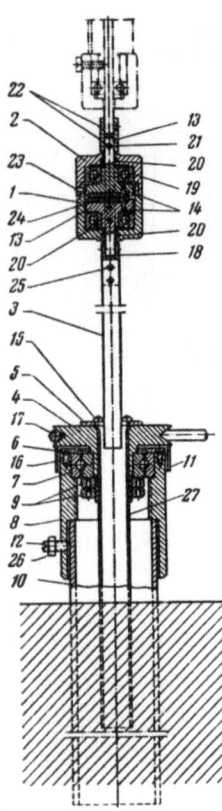

Fig. 24. The drive system: 1) lower half of clutch; 2) upper half; 3) drive rod; 4) pulley; 5) guide; 6) cover; 7) ball bearing; 8) body; 9, 26) nuts; 10) 27) tubes; 11) jacket; 12, 14, 15, 16, 23, 25) screws; 13) disc; 17) belt; 18) lower clutch; 19) radial ball bearing; 20) tilt adjuster; 21) upper clutch; 22) dowels; 24) spacer.

Fig. 25. The manual drive.

The Valve System

There are usually two valves, one each for the hydrogen and oxygen. There is also a stopcock in each line, which is used only to shut off the gas. This system does not provide accurate control of the gas flow, i. e., of the temperature. The temperature cannot be reproduced accurately in successive runs, and so it is difficult to produce a standard product. The system is inflexible, and cannot be used to shut off both gases rapidly and simultaneously. The result is to increase the number of boules with cracked or fused tops.

It is also impossible to shut the viewing port and cut off the gases simultaneously at the end of a run. The two operations are separated in time, which is the cause of the cracking or fusion in the crystals, because the temperature has time to change.

The valve system to be described provides smooth and accurate temperature control. There are two pairs of valves (Fig.9), namely B_3 and B_4 for the oxygen (see Fig. 27 for detailed drawings), which have micrometer adjustments 31 and 32, and B_1 and B_2 for the hydrogen; there is also a semiautomatic valve unit (see Fig. 28 for details). Valves B_2 and B_4 serve mainly to turn on the gases. At the end of a run they are closed very soon after the valve unit has cut off the gases. Valves B_1 and B_3 are the control valves proper.

If the gases are of constant composition and pressure, the valves can be given fixed settings, which may be left the same between runs designed to produce identical crystals.

In boule-growing, the micrometer adjustments 31 and 32 (Fig. 28) are fitted to the oxygen valve B_4, which has the function of varying the oxygen supply as the crystal is expanded (Fig. 9). The valve is opened slowly with the micrometer, either automatically or by hand. The micrometer screw 31 allows the flow to increase very slowly and smoothly. The crystal is allowed to grow freely when it has reached the preset size. The valve B_4 is then in its second standard position.

The valves are easy to take apart for repair, because the body need not be removed from the plant. The parts that wear the most rapidly, such as the nut and the spindle tip, are demountable and are easily replaced. The screw 31 operates via the worm gear 32 fixed to the spindle of the valve and sets the temperature very exactly. For rough setting the screw 31 is disengaged from the gear 32 by using the knob 33.

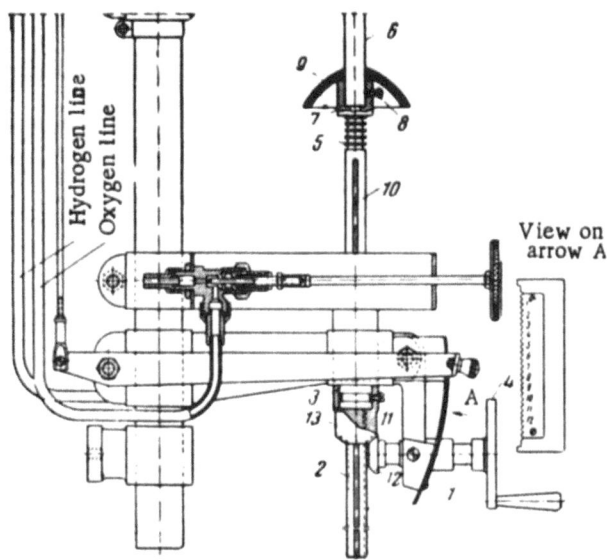

Fig. 26. The SP-3 manual drive: 1) axle; 2) lead screw; 3) guide bush;
4) wheel; 5) fins; 6) refractory candle; 7) candle holder; 8) grub screw;
9) ceramic skirt; 10) retaining slot; 11) stop screw; 12) cone gear; 13) compound bush.

The crystal is prevented from running at shutdown by the valve unit, which ensures also that the two gases are cut off instantly and simultaneously. The unit can be used with a single apparatus, or with a bank of them. The unit consists of two independent valves, which are coupled via a control system to the control lever 29 (Fig. 28). The valves have rubber membranes, which are pressed down by discs.

These membrane valves are sufficiently gas-tight and need not be made very accurately, which makes them easy to produce and use. The system is convenient and precise in operation.

This semiautomatic unit works as follows. Before the crystallization is started, or as soon as one group has been shut down completely, the unit is opened by hand. For this purpose the lock 25 is released. This frees the knob 29, which is pushed out by the spring 11, and allows the valve springs to expand fully; the rubber membranes uncover the holes 15. The valve is then in the working position. The unit is shut down at the end of a crystallization. The knob 29 is pressed rapidly home until the lock 25 is operated by the spring 23; the two gases are cut off simultaneously. The life of the valves is extended by opening them again a few minutes later, when the main valves have been closed.

The Automatic Crystal Expander

Crystal expansion is one of the most important operations in growing boules.

A very thin seed (some 2-3 mm in diameter) has to be built up to a diameter of 24-25 mm in a definite time. The temperature is raised slowly and smoothly by increasing the oxygen flow.

In the production plant this operation is done manually with the microvalve, and often results in scrap.

The flow resistance of the pipes, and the behavior of the valves, is not the same even in nominally identical units; the jets are unequal in size, the burners differ in resistance, and the furnaces vary in heat loss. Each unit has therefore to have its own calibration curve.

Now, the furnace temperature lags behind any increase in oxygen flow, and so one cannot see at once whether the oxygen flow is set correctly. The seed melts (perhaps right away) if the oxygen is let in too fast at the start. The process has then to be restarted. Cases are not rare in which a crystallization has had to be started 3 or 4 times. This operation is one of the most important ones, and so we have developed a simple device to do it automatically (Fig. 29). The oxygen supply and expansion time can be varied widely. Long use has shown that the device works exactly and reliably and is easily repaired. The system has much to commend it.

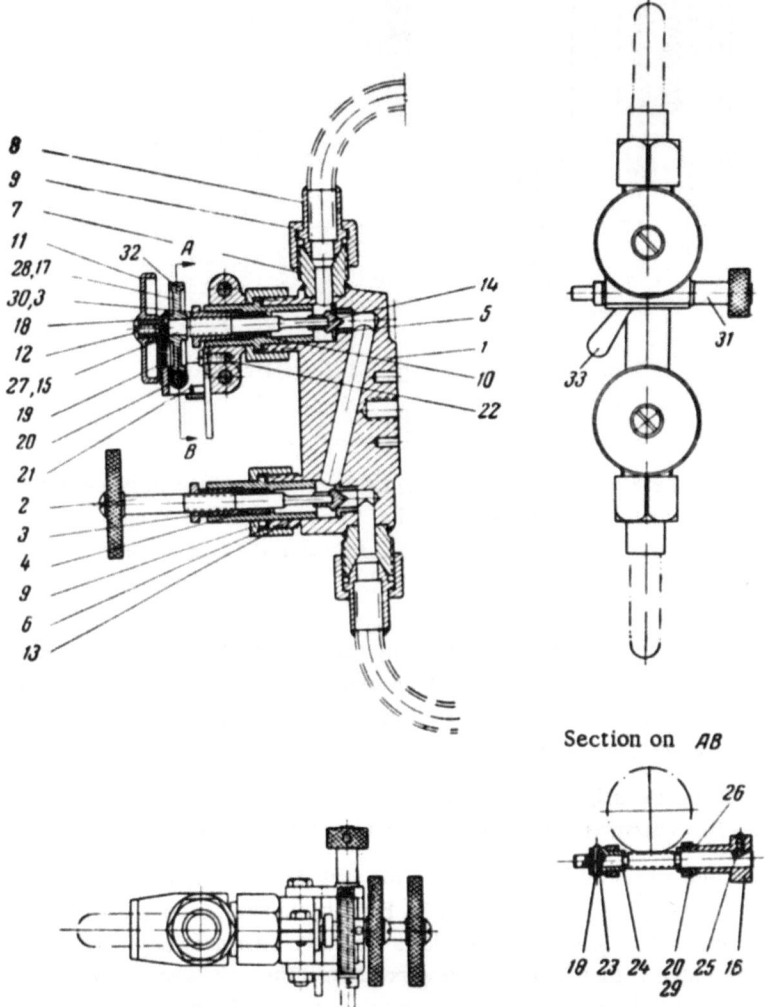

Fig. 27. The oxygen microvalve: 1) body; 2, 15, 27) valve spindles; 3, 9, 30) nuts; 4) asbestos packing; 5) seat; 6) insert; 7) inlet tube; 8) nipple; 10) packing; 11) wheel; 12, 25) screws; 13) vessel; 14)valve cone; 16) head of worm; 17, 28) gear wheels; 18) fastening; 19) spring; 20, 29) worm; 21) latch; 22) collar; 23) clamping ring; 24, 26) bushes; 31) micrometer screw; 32) worm gear; 33) knob for cutting out micrometer screw.

The expander is controlled in terms of the number of teeth on a ratchet wheel exposed to a dog. Figure 30 shows the kinematic system. The expander mechanism is coupled through a linkage to the micrometer valve. The mechanism is fixed to the valve stand. It can be cut out at any time.

The temperature in the furnace is determined by the gas flow. Two water gauges (see Figs. 32 and 7) are fitted, one for each gas, to record the gas pressures and flows during a run. The readings are made more accurate by inserting extra calibrated resistance units in the oxygen line (at the inlet in the head) and in the hydrogen line (at the burner inlet). The resistance units are such as to give readings of about $^2/_3$ full-scale at the maximum flow rates. Sliding verniers are used to read the menisci accurately. The heights of the water columns are set in accordance with the gas pressure by using these sliding scales.

Plexiglas rotameters are often used to maintain the furnace temperature, but they read the gas flows only roughly. They wear rapidly and the readings change, and so they are unsuitable for the purpose.

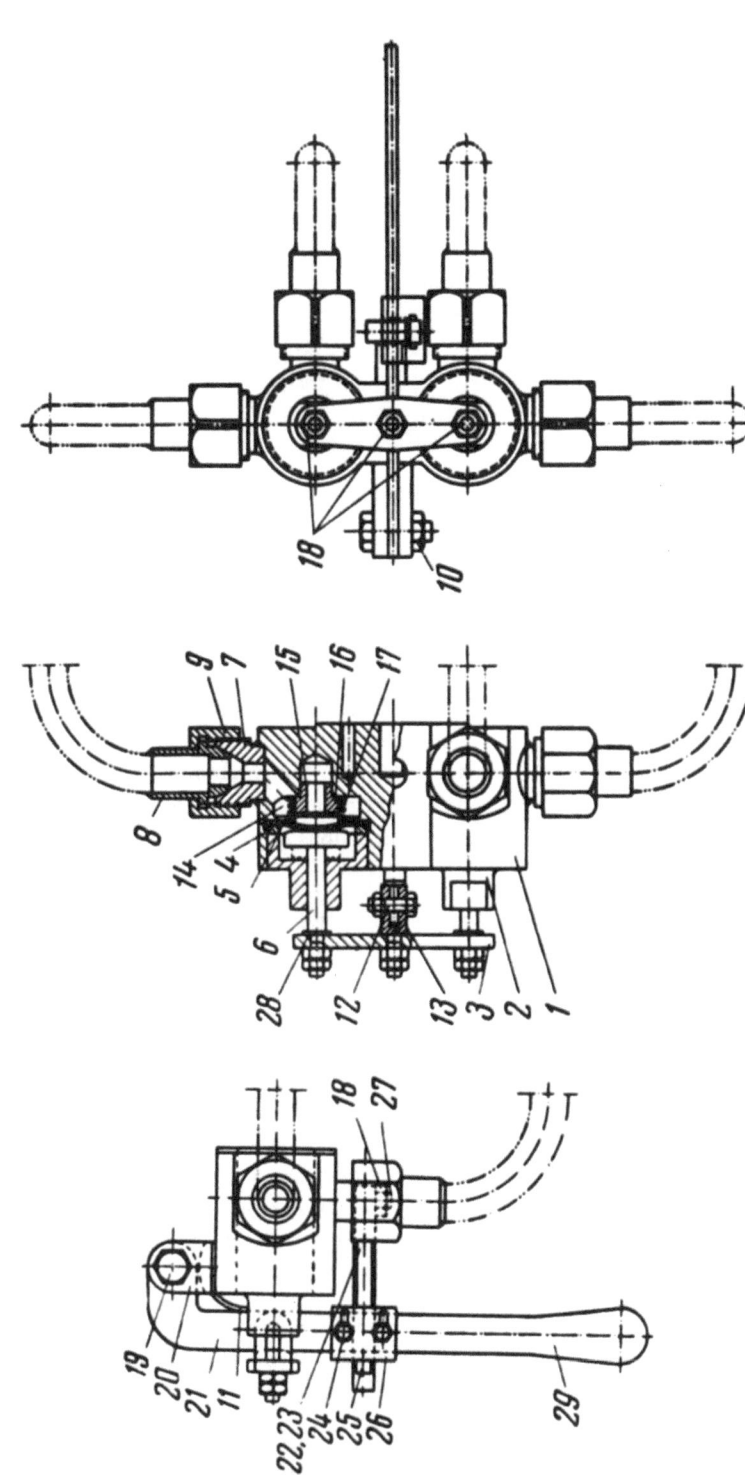

Fig. 28. The valve unit: 1) body; 2) valve cover; 3) coupler; 4) pressure ring; 5) diaphragm; 6) valve system; 7) inlet tube; 8) nipple; 9) pressure nut; 10, 18) nuts; 11, 17, 23) springs; 12) bolt; 13, 20) lugs; 19, 24) bolts; 14) ring; 15) seat; 16) insert; 21) lever; 22) screw; 25) lock; 26) axle frame; 27) axle; 28) washer; 29) control knob.

Fig. 29. The automatic crystal expander.

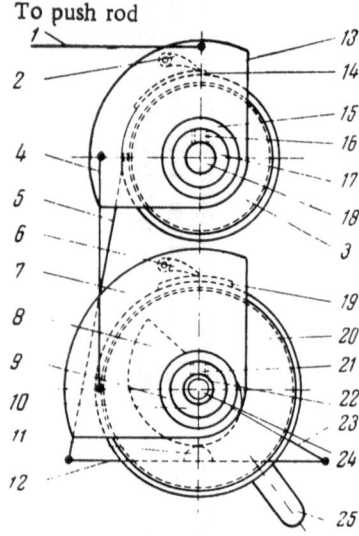

Fig. 30. The kinematic system of the automatic expander: 1, 4, 5) links; 2, 6) dogs; 3, 20) ratchet wheels; 7, 13) plates; 8) cam; 9, 15) nuts; 10, 17) knobs; 11) follower; 12) lever; 14, 19)covers; 16) grub screw; 18) worm; 21) clamping screw; 22) groove; 23) needle; 24) axle; 25) lever.

Assembly and Adjustment of the Main Units

The improved apparatus was designed to be set up on a framework fastened securely to the ground, with the units held on special arms (Figs. 9 and 10), to ensure that the apparatus was easily and rapidly assembled, adjusted and serviced. If it is absolutely essential, the crystal drive need not be mounted on the common frame, but only if these conditions are complied with:

1) the crystal holder is coaxial with the burner at all settings, and

2) the vibration is very slight.

The apparatus can be set up anywhere on the framework 7 (which is made of rustproof material), which has two horizontal rods running the full length and rods, at 90° to these, set in the ground. The whole is made rigid by sloping tie-bars to the ground from the top of the frame. The tapper is mounted on arm 8, the head on arm 9, the furnace on arm 10, the automatic crystal drive on arm 43, and the entire valve system on arm 11.

Crystallization of Rod Corundum

We deal with rod corundum because this form of the material is still fairly new.

The seed 29 or candle (Fig. 8) is first set exactly at the center by using the viewer 44 and the adjustable holder 45. The settings of the automatic devices, of the valves, and of the charge-feed levers are then checked. Unit 24 must be switched off, valve B_4 must be tightly closed, valve B_3 must be set to the required position, and the valve unit must be in the open position. Lever 2 must be in the working position, and lever 3 in the idling position, i. e., must cut off the charge feed. Then an open flame on the end of a long rod is brought as close as possible to the viewing port (without disturbing the viewer), and the hydrogen valve B_2 is opened slowly (with the head turned away from the port, to avoid burns). When the hydrogen has lit (sometimes with a slight bang) the flame is withdrawn, and the furnace is allowed to warm up for a little while. Then the oxygen valve B_4 is opened slowly until the hydrogen flame no longer lights up the inside of the furnace. There is often a slight pop at this moment.

The oxygen valve should never be opened first, or an explosion may result. As soon as the oxygen is admitted the hydrogen valve B_2 is opened fully, and the oxygen valve B_4 is still opened slowly, in order not to heat the lining up too rapidly. The furnace is left for 30-40 min to reach a constant temperature.

If several units are to be started at once, the hydrogen is lit in each by turns. Then the oxygen is turned on in each unit in turn, as for one unit. These operations can be carried out more quickly if the furnaces are already hot. When the set temperature has been reached the candle is inserted slowly until it lies a little below the viewer. The automatic drive to the tapper and candle is cut in only when the candle has had time to warm up. Then the charge lever 3 is operated. Lever 3 must be set in the working position, which has been set previously. Levers 2 and 3 may need to be adjusted at the start of a run if the composition or quality of the charge, or the crystallization conditions, have been altered.

A cone of sintered powder first forms on the candle. This cone is watched to see if it is regular in shape, with a sharp apex. The charge should fall on a small area, and should not scatter. The apex enters a hotter zone as the cone grows, melts, and forms a small crystal nucleus as a thin knife-edge. The thinner this edge, the less often it cracks, and the better the chances of producing a monocrystal. The automatic crystal drive is started when the top of this knife-edge reaches the level of the viewer (which implies a length of about 3.5 mm). To do this the handle 25 (Fig. 8) is used to set the dial 24 in the usual position.

The top of the rod must be watched throughout the crystallization. The top should always be at the same level as the viewer. The charge flow has to be adjusted with levers 3 and 2 in turn if the rod grows too quickly or too slowly. When this adjustment has been made, the speed of the drive is set to keep the rod at the proper level. The speed is increased if the rod has risen above the viewer; to do this the dial 24 is turned a few divisions to the left. If, on the other hand, the top has fallen below the viewer, the dial 24 is turned a few divisions to the right. The dial 24 is reset to the original position only when the top of the rod has returned to exactly the right position. Sudden movements are not allowed.

All adjustments must be made smoothly. Any sudden disturbance damages the crystal. One of the main signs of disturbances, or of inexact regulation, is the appearance of streaks on the surface.

The crystallization is stopped when the rod has grown to its set length (i. e., when the lower end emerges from the furnace). There must be no cold air currents near the plant at this time.

The charge feed is first cut off with the lever 3, and then the rod is lowered a little (3-5 mm). The port is then closed with the slide 33 at the same time as the gases are cut off by operating the valve unit; then valves B_4 and B_2 are closed. The crystal drive is turned off 15-20 min later, together with all the rest of the equipment. The rod is then wound out by hand. It is gripped with tongs having asbestos inserts and removed slowly and carefully from the holder.

The operations are done in the same sequence with each unit if a bank of units is shut down. The valve unit is set to the open position when the bank (or single plant) has been closed down completely. The pressures of the gases must be watched continuously on each unit, or on a group of units. The gases must also be checked regularly for composition by analysis. The crystallization is stopped if the pressure or composition fluctuates beyond acceptable limits. Trial runs are done with different speeds to find the best range for use with a given charge batch; the best crystals are made in this way.

Another gas control system may be used if it is found that the tops of crystals run, when several out of a bank of units are closed down (because the gas pressure rises). The separate valve units (one for each plant) are replaced by a common unit. In this system the start and stop operations on each individual plant are run in the following sequence. First, the charge flow is cut off. The port is closed and the rod (or boule) is lowered a little, both at the same time, in turn for each plant. Then the common gas supply is cut off, and then the B_2 and B_4 valves are closed. All the subsequent operations are done in the sequence given above.

The technical regulations in force with the USSR require that only dark ruby stones be used for watch parts. The greater the chromium content the more difficult it is to crystallize the material. The size of the crystals falls as the chromium content increases, and the rate of growth decreases; the internal stresses are larger, and so the rods often crack. We were unable to make very dark rubies until the present apparatus was produced. It is now equally easy to make colorless corundum and dark rubies.

The properties of parts made from synthetic corundum depend very much on the orientation of the stone. It is thus important to be able to grow oriented rods. The process differs from the ordinary one in which seeds are used. It is essential to use conditions such that a polycrystalline mass cannot form on the seed. To do this the seed is put into a preheated furnace, is set into position, and is slowly brought up to the working temperature. When a steady state is reached (the time required is found by trial in each case) and the top of the seed has melted and broadened, a little, the feed is turned on and gradually brought up to the normal rate.

If the top does not broaden in this position, the seed is lifted up a little until it does; the crystallization is then started. The top is set opposite the viewer when the seed has grown by 2-3 mm, and the subsequent growth is as described above.

Boules differ from rods in that the seed has to be expanded, and in that the free growth-rate in the cylindrical section is lower.

SUMMARY

In January 1950, 26 sets of the improved apparatus were put into use.

The first experimental batch of dark rubies was passed to the stone-working factory, and was found to differ appreciably from the usual Soviet and foreign materials. The crystals were longer and broader than usual. They had spherical tops, true cylindrical shapes, and the sides were coated with a rippled deposit of charge.

The crystals were unusually resistant to cracking. Some boules were sawed up into plates, and none cracked, not even the dark ruby ones; intact plates of full diameter were produced.

These results were obtained with routine production charge and gases, which confirmed the author's view that the main factor controlling crystal quality was the crystallization conditions, and that the correct technique and improved apparatus would improve the quality of the rubies considerably.

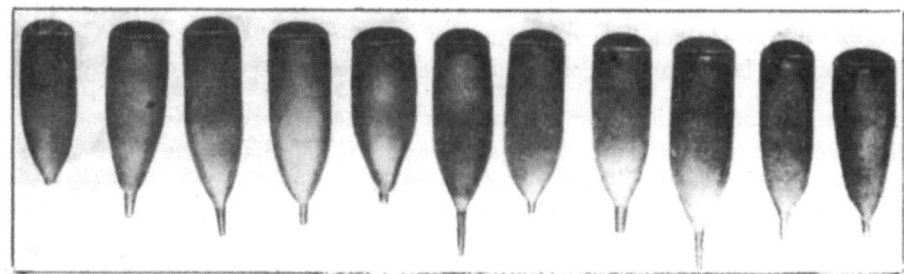

Fig. 31. Ruby crystals grown on the pilot apparatus (reduced by a factor 1.5).

The apparatus is easy to control and is stable in characteristics; the boules are constant in size and regular in shape (Fig. 31).

The apparatus has been found satisfactory for use in making dark rubies.

Pure electrolytic gases have to be used in growing rods.

APPENDIX

Instructions for Setting Up and Adjusting the SP-3-1939 Apparatus

The dimensions of certain parts may vary. These parts are the oxygen tube, the mixing chamber, the furnace, the candle, the tapper, etc. It is therefore not of much value to give exact distances between parts; the

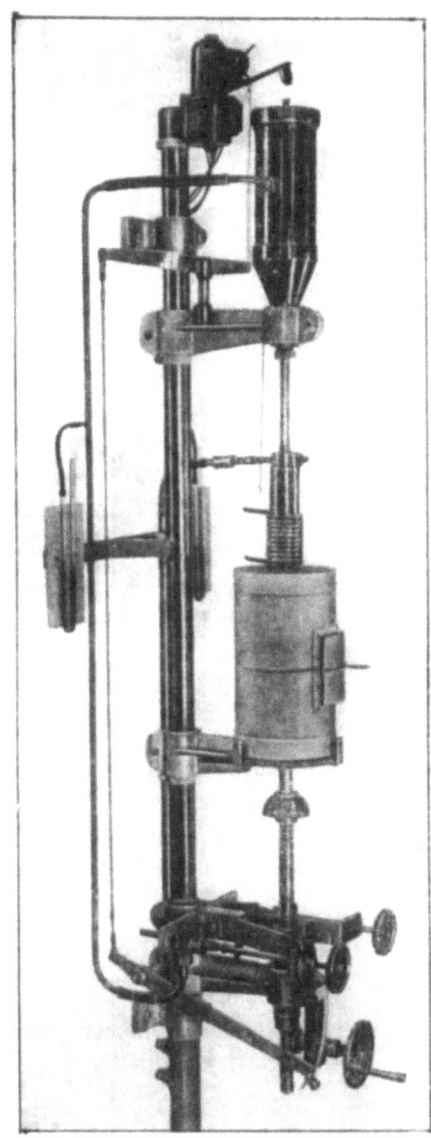

Fig. 32. The SP-3 crystallization apparatus.

following steps are best used to ensure that the apparatus is set up quickly and accurately.*

The apparatus (Figs. 32 and 33) is set up in the following sequence.

1. The support arms 7 and rod 8 are bolted to the frame or wall.

2. The rod is set to be vertical in two planes by using the 90° lines engraved on it; packing is inserted under one arm for this purpose.

3. The rod is slackened in its bosses on the arms and is lowered until the top end comes clear of the boss. The following parts (previously set ready to hand) are then slid onto the rod: the boss for the crystal drive 9; the boss for the microvalves 11; the supporting ring 10; the furnace arm 12; and the boss that carries the arm 13 for the head and burner. Then the rod is replaced in the top boss and is lifted up to the right level; the bosses are then tightened up.

4. The bosses are slackened roughly as shown in Figs. 32 and 33, and the head with the shaken box 2 and burner 3 is fixed to the top arm.

5. The furnace arm is checked for level; then the furnace 4 is mounted.

6. The tapper mechanism 1 is mounted upon a ring 10 placed at the top of the rod.

7. The arm 13, with head and burner, is slid along the rod until the closure 14 touches the tapper 15, when the latter is horizontal. The boss is then tightened.

8. The burner 3 is fixed temporarily to the oxygen tube 16 in the extreme "out" position. Then the arm 12, which carries the furnace, is lifted up until the burner enters the furnace and touches the cylinder 17. The boss and support ring 10 are lightened and the burner is slackened off.

9. Then the arm 9, which carries the crystal holder, is set. The spindle 18 is lowered as far down as possible, and a candle 6 is set up in the candle-holder 20. The arm is set in a position such that the furnace support and the end of the candle are far enough apart for the finished crystal to be removed; the boss is then tightened.

10. The valve arm 11 is lowered onto the crystal-holder arm. Arm 11 is not needed if the valves are fixed to the frame.

The mounting is completed by this last step, and the centering is begun.

The tapper mechanism is turned to one side every time the shaker box is refilled, so before each crystallization the tapper 15 has to be centered, to the closure 14.

The other parts are centered up in the following order.

---

* The description implies that all the parts are assembled on a common vertical rod, as are parts on an optical bench — Compiler.

133

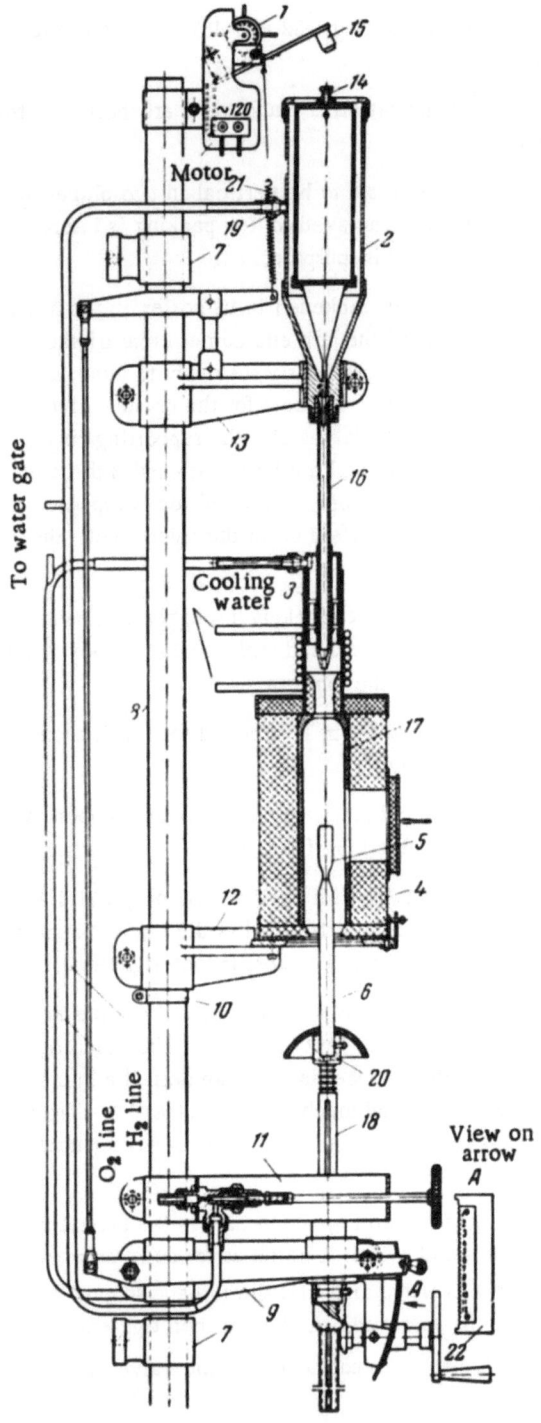

To water gate

Motor

Cooling water

O₂ line

H₂ line

View on arrow A

Fig. 33. The SP-3 crystallization apparatus: 1) tapper mechanism; 2) shaker box; 3) burner; 4) furnace; 5) crystal; 6) candle; 7) boss arm; 8) rod; 9) crystal-holder arm; 10) support ring; 11) arm with valves; 12) furnace arm; 13) head arm; 14) closure; 15) tapper; 16) oxygen tube; 17) cylinder; 18) spindle; 19) inlet tube; 20) candle holder; 21) spring; 22) scale.

1. The burner 3 is demounted. The candle 6 is removed. The furnace 4 is swung aside on its arm. The head and shaker box 2 are set up on the arm in such a way that the inlet tube 19 points to the left; they are then clamped. A nut and plumb bob are screwed on at the point where the burner is fixed, and a center is moved about in the horizontal plane until it is lined up exactly to the bob, after which the arm is clamped. The candle is then set in place of the center and is lined up to the bob; it is then clamped.

2. The bob is removed. The oxygen tube 16 is fixed on and is centered to a bob hung from the base of the head cone by bending it in two planes at 90° to one another. The body of the burner is then fixed onto the oxygen tube and is held up by ties to the head (Fig. 33). The head cone and oxygen tube are centered into the same plane against a scale on the support arm. The centering need not be repeated when the head or burner is demounted or replaced.

3. The furnace is swung on the rod 8 to come into position above the refractory candle 6, which is brought up gently into the inside of the furnace; the furnace is centered up to the candle with three screws, and the boss is clamped.

4. The burner is released from the ties and is lowered onto the furnace; the setting is correct if the burner enters the furnace exactly. If it does not enter properly, the furnace is centered up to the burner by inserting iron strips under its base.

5. The arm 11 with the microvalves is centered by eye to the crystal drive and is then clamped.

This completes the centering. The burner 3 is finally centered during a trial crystallization.

The gas and charge-feed devices are then coupled up. These operations are simple and are made clear by Fig. 33.

The choice of spring 21 in the feed is important. The strength must correspond to the working part of the scale 22 and to the strength of the opposing spring. If the knob on the control lever is set at the top mark, the tapper should not strike the closure at all. With the knob at the bottom mark the blows should be such as to supply more charge than is needed.

All the joints in the gas lines must be checked for tightness. Soap solution is used to test the various parts for leakage.

The automatic drives are assembled on arm 11 and on the crystal drive arm 9. They could also be set up on the framework or on a stand.

The mechanism for withdrawing the holder automatically is mounted entirely on the drive arm (Fig. 20), except for the motor, which is fastened to the wall or floor on a special support.

The automatic crystal expander is assembled entirely on the microvalve arm (Fig. 29).

The apparatus can be fed from a low-pressure gas line via pressure regulators, or from cylinders via reducing valves and regulators. In both cases the gases are brought to the valves in copper pipes or armored flexibles, either directly or via blowback valves.

The regulators (for low pressures) and reducing valves (for high pressures) must work without sticking and must hold the pressure constant, otherwise the furnace temperature will fluctuate and the crystallization will be spoiled.

LITERATURE CITED

[1] F. Kraus, Synthetische Edelsteine (Berlin, 1929).

[2] M. Gaudin, Compt. rend. 4, 999 (1937).

[3] H. Senarmont, Compt. rend. 32, 410 (1851).

[4] J. Ebelman, Chem. Phys. 22, 213 (1848).

[5] E. Fremy and E. Feil, Compt. rend. 85, 1029 (1877).

[6] E. Fremy and A. Verneuil, Compt. rend. 104, 738 (1887); 106, 565 (1888).

[7] P. Chirvinskii, Artifical Production of Minerals in the Nineteenth Century [in Russian] (Kiev, 1903-6).

[8] S. K. Popov, Izv. AN SSSR, Ser. Fiz. 10, 5/6, 505-9 (1947).

II. USE OF CORUNDUM ROD IN MAKING ARTIFICIAL FIBERS

(Work done in collaboration with A. A. Popova)

INTRODUCTION

Fiber guides are rods used to guide and stretch fibers during formation.

The guide material must be hard enough to work for long periods, and must take a good polish, in order to give high-quality fiber. It must be stable to the reagents used in making the fibers. Synthetic monocrystalline corundum satisfies all these demands.

Until recently, however, corundum rods have not been used for the purpose in the USSR, because the consumption of cutting diamonds in forming the rods from boules and in polishing them would have been excessive.

Our method of making corundum rods, which was designed to supply the watch and instrument industry, enabled us to propose that corundum should be used extensively for fiber guides.[*] A method and apparatus for flame-polishing the rods, in which diamonds were not used, was developed; so was a method of bending the rods automatically.[**]

Our technique of making corundum guides is described below, together with the apparatus used to make the guides on a large scale. Some special holders for the guides are also described.

---

[*] S. K. Popov, Author's Certificate No. 85910 (Fiber Guide) of June 20, 1950, with priority from April 22, 1949.
[**] S. K. Popov, Author's Certificate No. 103103 (Device for Bending Corundum and Similar Materials) of May 8, 1956, with priority from October 16, 1953.

Glass, porcelain and agate have until recently been the main substances used for fiber guides. Attempts have also been made to use plastics and ceramics.

Glass and porcelain guides are the main ones used for viscose fiber, but they wear rapidly; glass ones wear out in 4-6 to 4-5 days, depending on the type of fiber, because after this time the surfaces have become so scored (Fig. 34) that the elementary strands from which the fiber is drawn begin to get damaged.

Fig. 34. Glass guide after 24 hours. × 70.

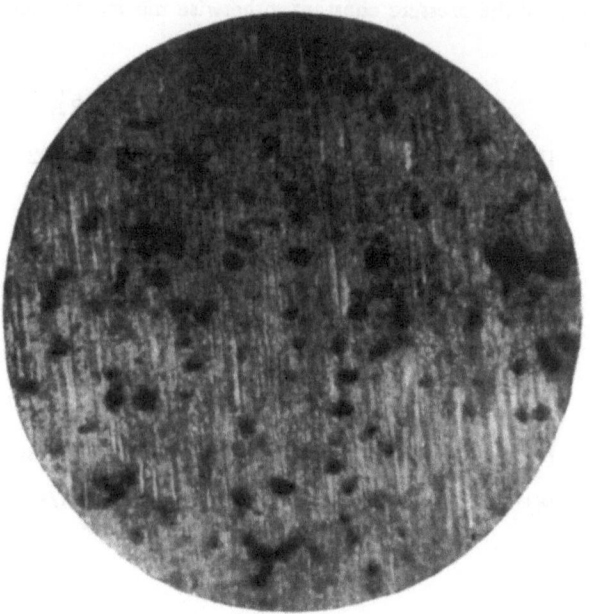

Fig. 35. Porcelain guide, unused. × 70.

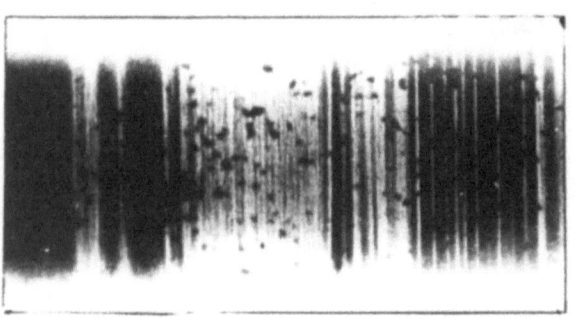

Fig. 36. The same surface after three days' work. × 70.

Porcelain guides (Fig. 35) show hairline scratches of various sizes. These scratches rapidly enlarge into deep grooves (Fig. 36), which start to break the fibers, and the guides have to be scrapped. Such guides last for 4-6 days with matt silk, and up to a month with sheer silk.

Agate guides are used mainly with capron fiber, and not very much with viscose fiber. They last for 2-3 months with capron, but for not more than a month with viscose. Figure 37 shows the polished surface of an agate guide seen under high magnification. The surface is greatly changed by use (Fig. 38). The grooves have sharp edges, which at first damage the strands, and then break them. The surface is repolished with a diamond, whereupon the guide can be reused.

The main disadvantage of agate is that diamonds, which are in short supply, have to be used to polish the rods.

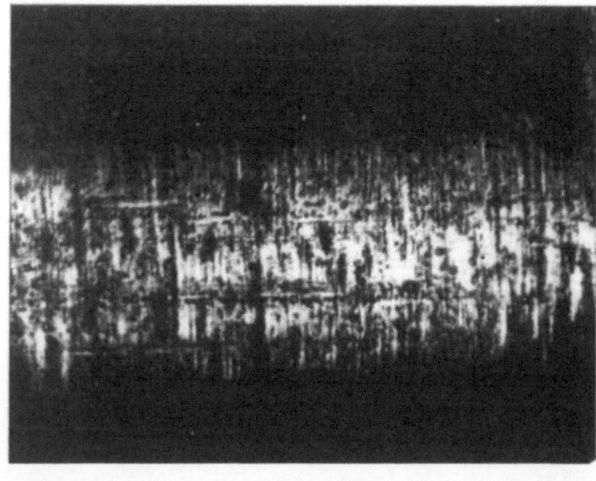

Fig. 37. The surface of an unused agate guide. × 70.

Corundum is free from the faults of glass, porcelain, and agate. Constant observation on corundum guides that have been in continuous use for years shows (Fig. 39) that

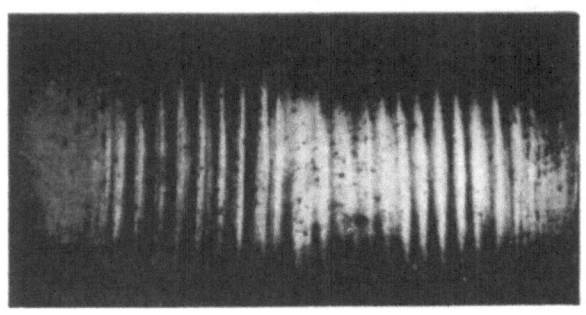 

Fig. 38. The same surface after two months' work. × 70.

Fig. 39. Surface of a corundum guide after nine months' work. × 70.

the fibers do not score the surface and produce sharp-edged grooves – in fact, they polish it. This self-polishing effect appears within the first year of use, and is one of the remarkable properties of the material.

It is found that corundum guides become much hotter than agate ones in forming capron cord. The cause of this effect has not yet been discovered.

Guide Manufacture

The following are the main steps in making corundum guides by our method:

1) making the corundum rod,

2) polishing the rod,

3) bending the rod,

4) cutting and finishing the guide.

Making Rod Corundum

The apparatus and methods for making rod corundum are dealt with in the preceding paper.

Colorless crystals 4-4.5 mm in diameter are best used for guides. Diameters over 4.5 mm are undesirable. Rods of large size have internal stresses such that they crack either during manufacture (especially during cutting and grinding) or during use.

Rods to be used for guides should show neither unfused rims (which reduce the strength nor bubbles larger than 0.1 mm near the surface. Such bubbles may cause sharp-edged pits during polishing. Cloudy or bubbly regions near the center are not objectionable. Rods for use as guides may be grown using the same gases as are used for boules.

Corundum rods are sometimes elliptical to the extent of 0.1 mm or more. This fault should not occur if the apparatus is operated properly and the gas supply is steady. Rods of ellipticity not exceeding 0.1 mm can be used for guides.

Polishing Corundum Rods

The corundum rods were at first polished in the usual ways, i. e., with chemical agents. Corundum dissolves readily in certain fused substances at high temperatures. We have tested $KHSO_4$, $K_2B_4O_7$, $Na_2B_4O_7$, cryolite, and combinations of these. Borax and cryolite dissolve corundum the most readily. The best for polishing purposes is borax; cryolite does not give a smooth surface.

Polishing tests were done at 760, 800, 820, 850, 900, 950, 960, 1000, 1060, and 1100°C, with times varying from 5 min to 1.5 hrs. Temperatures below 800°C were ineffective; the surfaces stayed matt. A gradual rise in

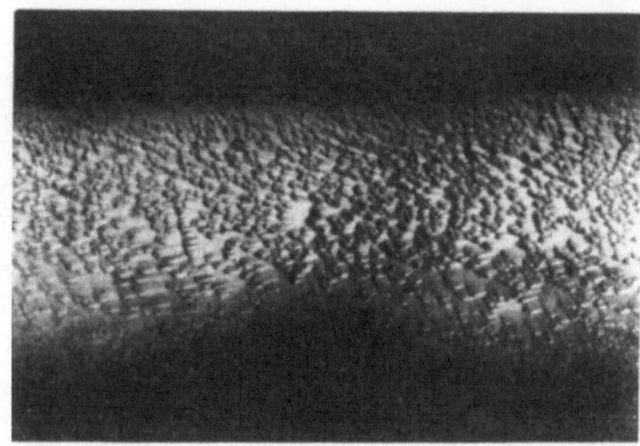

Fig. 40. Etched surface on a rod polished with borax at 1000°C for an hour. × 70.

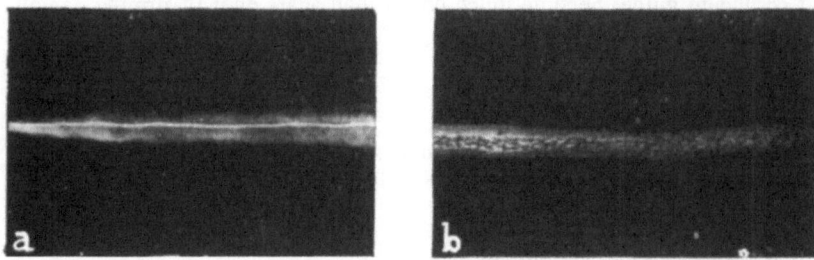

Fig. 41. Narrow straight "etch" bands on a rod polished with fused borax. × 20.

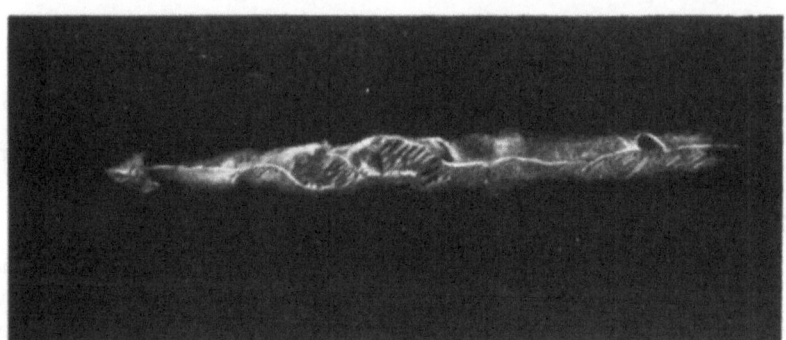

Fig. 42. Wavy "etch" band on a rod polished with fused borax. × 20.

temperature to 950°C also failed to give the desired result, although the solubility of corundum in borax rose appreciably. At 960°C we obtained a smooth, evenly polished surface. The treatment time was reduced to 30-40 min at 1000-1100°C, and the polishing was better. However, the metal parts of the polisher were attacked very rapidly above 1000°C. It is not altogether correct to say that the surfaces produced at temperatures above 960°C were smoothly polished, because, in fact, they showed more or less pronounced longitudinal bands [1]. These bands differed in type in accordance with the degree of polishing; rods not completely polished showed broad matt bands visible to the unaided eye. Figure 40 shows such bands at × 70.

Smooth wavy bands running along the length (Fig. 41, a and b; Fig. 42) were produced by borax and other powerful solvents. The spaces between the bands were clean and free from defects. Rods with such bands appear smooth to the unaided eye (Fig. 43).

The bands sometimes lie in the plane passing through the geometrical and optic axes of the crystal, but this is by no means always so, and the bands can occur at any angle to this plane.

Rods polished in the above way were tested as guides with viscose, and were found to be quite satisfactory. They were sometimes unsatisfactory with capron fiber, which they damaged. We therefore had to look for other methods, which would remove the defects left by chemical polishing. In 1952 we developed a method of thermal (flame) polishing, which had substantial advantages over chemical polishing.

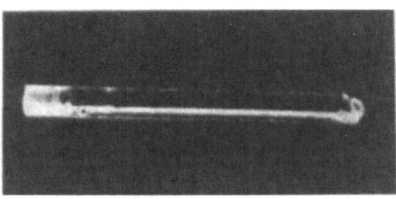

Fig. 43. General view of a thermally polished rod.

An oxyhydrogen flame is used for thermal polishing. The rod is preheated in a furnace, and is then withdrawn through the flame from the burner. The flame fuses the outer layer (especially the matt encrustation) and the rod becomes smooth and polished. The gas conditions are the same as for crystal growing. Rods polished at once after manufacture were found to have a better polish than ones that had been stored for a long time. The rods could not be polished actually during growth, because particles of charge tended to miss the growing surface and to fall on the polished sides. We found that the flame should come from the side during polishing, and not from above, as during crystallization. A special apparatus was designed for polishing the rod.

Apparatus for Flame-Polishing Corundum Rods

This apparatus (Figs. 44 and 45) consists of the following units:

1. The burner 1, which produces an oxyhydrogen flame with a central oxygen cone. The burner is inclined at 15-17° to the furnace. The jet 2 has a hole 3.5-4.5 mm in diameter; there is also a ceramic insert 40 mm long and 20-22 mm in diameter.

2. The muffle furnace 3, which is made of high-grade refractory. The muffle furnace ensures that the rod is appropriately preheated, and protects the polished crystal from the air. Normal crystallization furnaces can be used with rods 100-120 mm long; longer rods require furnaces up to 500 mm long, with working spaces up to 40 mm in diameter.

3. The holder 5, which withdraws and turns the crystal, and the refractory candle 4, on which the rod is fixed.

The best speed for use in polishing is 20-25 mm/min. The speed of rotation, as during crystallizations, is 110-115 rpm. Fixing the rod presents some difficulty. The rod is held most securely if the furnace aperture exceeds the diameter of the rod by only 0.1-0.2 mm. A gap of more than 0.2 mm allows the rod to waggle, which causes the surface to fuse unevenly, and so spoils the polish.

It is best to grade the rods into sizes and to make a candle for each batch with a recess exceeding the batch diameter by 0.1-0.2 mm. This speeds up the process and improves the polish.

The valve block is the same as that used for controlling the crystallization apparatus. The oxygen and hydrogen throttles are both 1.15 mm in diameter. The burner and gas-flow settings are found by trial for each unit. The approximate spacing from the burner should be 75-80 mm, with flow rates of 500-550 liters/hr (hydrogen) and 250-300 liters/hr(oxygen).

Pressure regulators and pressure and flow gauges are connected in both gas lines to facilitate control and adjustment.

The operations of lighting, heating and turning off are the same as for the crystallization plant, and so we shall deal only with the main points in the polishing process.

The rod is fixed firmly in the candle, is centered up, is set turning, and is inserted until the candle almost enters the flame. The top of the candle should be heated to redness. If the carborundum candle is not heated properly, it withdraws heat from the corundum (because it has a much larger thermal capacity), and so the end of the rod is not polished properly. The rod is withdrawn automatically as soon as the top of the candle is red hot.

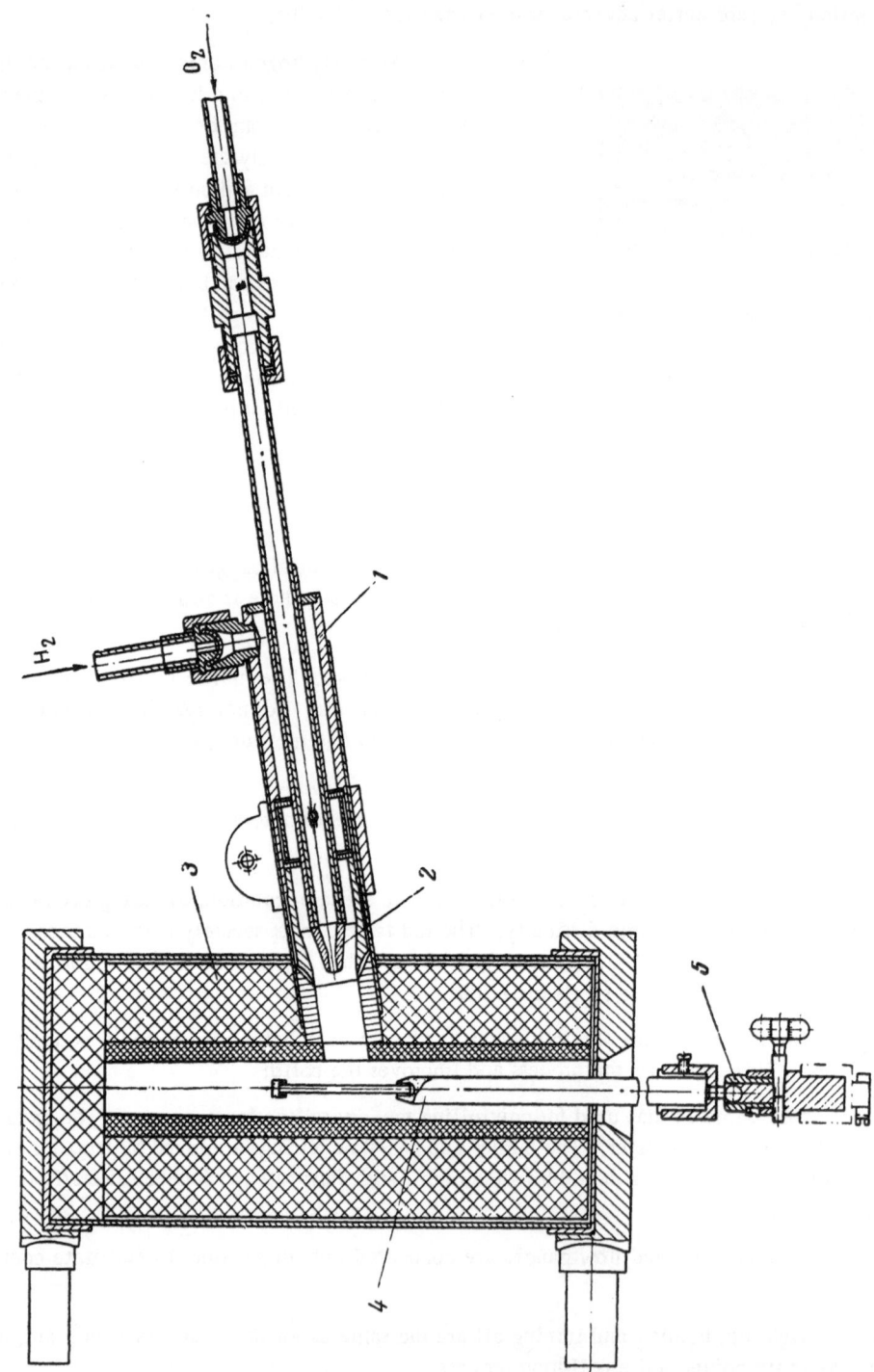

Fig. 44. The flame-polisher: 1) burner; 2) jet; 3) muffle furnace; 4) refractory candle; 5) holder.

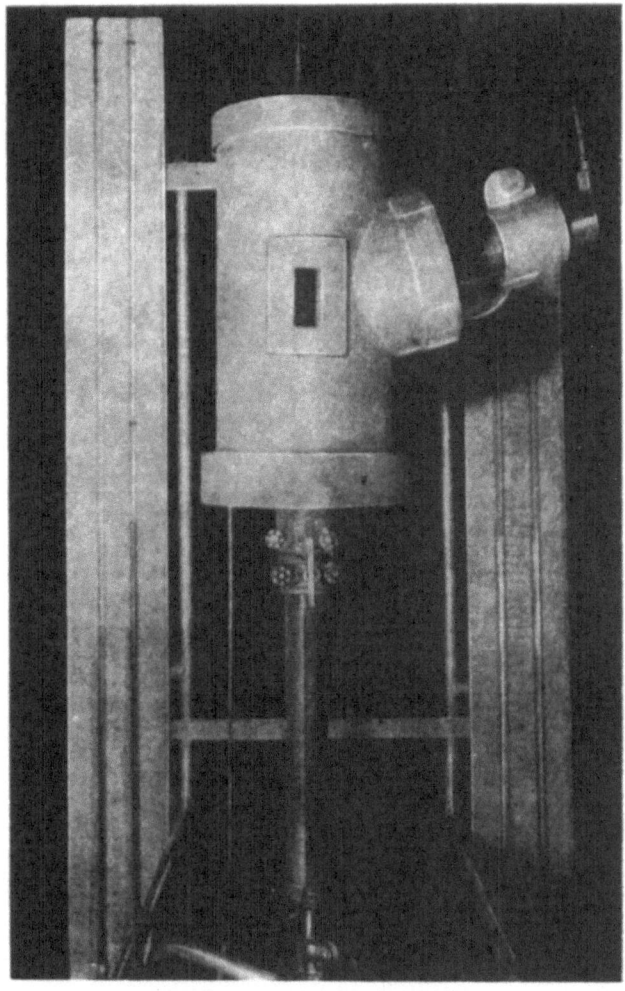

Fig. 45. External view of the rod polisher.

The following conditions are essential if the polishing is to be even:

1) the rod must be set exactly on the vertical axis of the furnace,

2) the gas flows must be constant,

3) the withdrawal rate must be constant,

4) the rotation speed must be constant.

The polish is not of the best if any one of these conditions is not complied with. The conditions should be observed most carefully.

We have polished many rods of differing diameters and orientations with various withdrawal rates and gas flows. The rods did not all behave in the same way under identical conditions.

Polishing Rods with Differing Crystallographic Orientations

In our polishing work we took account of the orientation of the rod axis relative to the optic axis by measuring the angle α between them. The orientation relative to the symmetry plane and to the twofold axis was random. We found that the variously oriented rods behaved differently under severe thermal stress. By severe stress we mean temperatures such as to convert a monocrystal withdrawn at 80-100 mm/min to a polycrystalline condition.

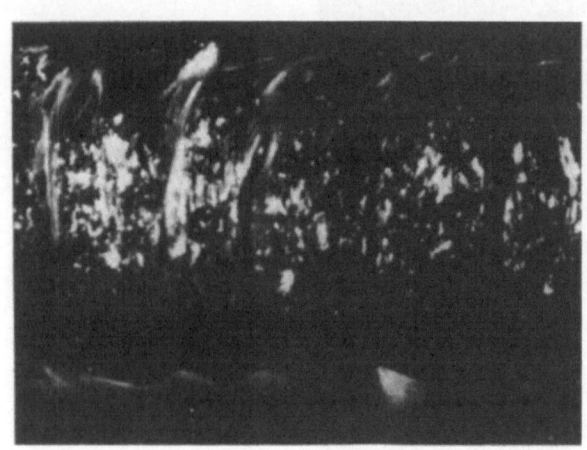

Fig. 46. Band produced by flame polishing on the surface of a rod with α between 80 and 90°. × 30.

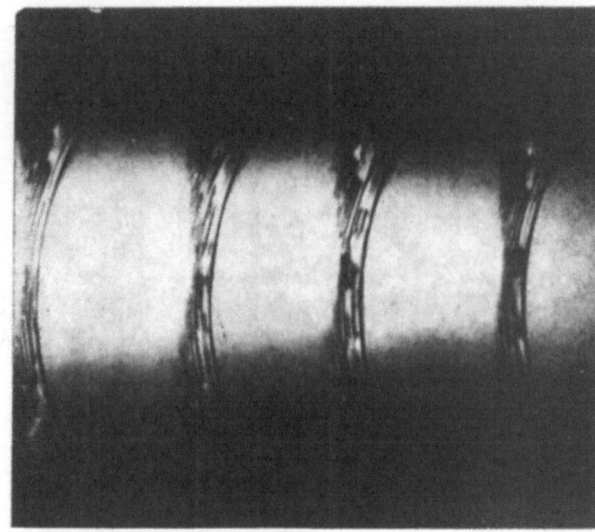

Fig. 47. Band produced by flame polishing on the surface of a rod with α between 70 and 80°. × 70.

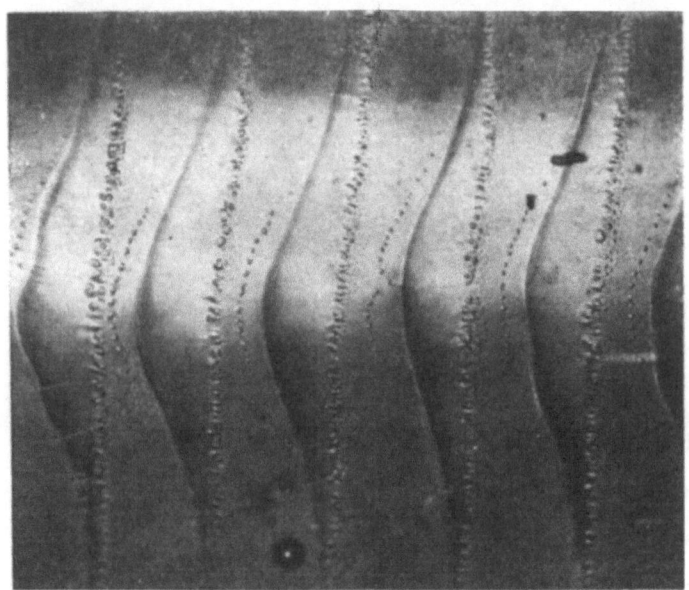

Fig. 48. Bands produced by flame polishing on the surface of a rod with α between 60 and 70°. × 70.

These conditions regularly produced two symmetrically placed longitudinal bands in crystals with α greater than 80°. Sometimes both bands extended right along the crystal; sometimes one could be seen only at the base (gripped end), or was discontinuous, whereas the other was always continuous.

The bands are made up of large sharp segments which are quite distinct from one another (Fig. 46). The bands can be seen easily with the unaided eye, and never appear with rods of other orientations. They appear only in the plane containing the optic and geometrical axes.

Crystals with α in the range from 70-80° give two longitudinal bands of a different type (Fig. 47). One is usually clearer and more symmetrical than the other. These bands always lie in the plane containing the optic and geometrical axes and can be seen under high magnification to consist of steps whose edges have a curved

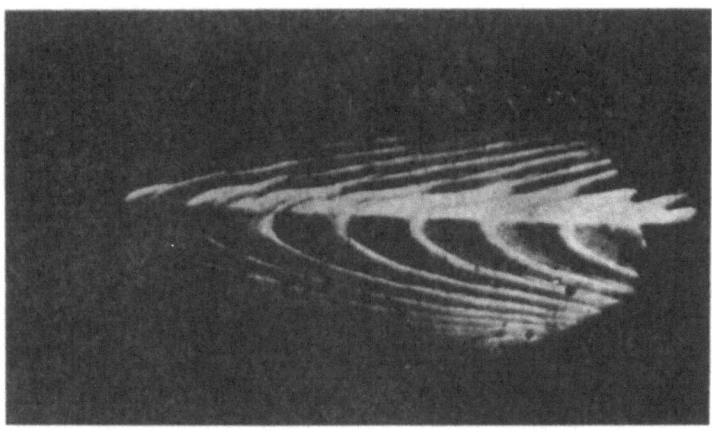

Fig. 49. Bands often produced by flame polishing. × 20.

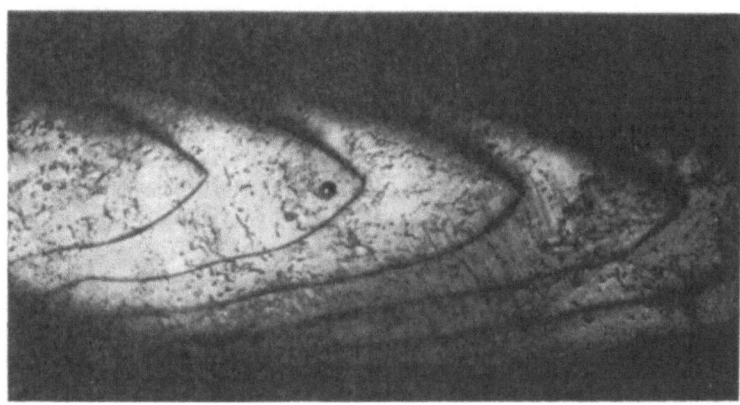

Fig. 50. The surface of Fig. 49. × 70.

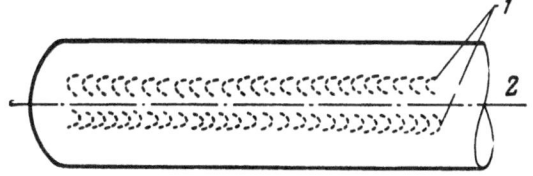

Fig. 51. Relation of the bands (1) shown in Fig. 52 to the plane (2) which passes through the growth and L_6^3 axes.

outline. The following regularity is notable. The largest arcs are equal, and of a length (in degree terms) roughly equal to the angle between the optic and growth axes.

Rods with angles of 60-70° usually give bands that lie in a plane normal to the plane of the optic and growth axes (Fig. 48).

The most frequent form seen on crystals polished under severe conditions is composed of drawn-out curves (Fig 49 and 50). These occur on rods with orientation angles of 50-80° and lie only in planes normal to those passing through the optic and growth axes.

Rods with $\alpha < 50°$ sometimes show bands of the same type, but these bands do not extend right along the rod; there may be six, four, or two, at the gripped end only. These rods usually show no specific bands at all, and the polishing defects appear as scattered hummocks joined by short lines.

Rods with these various types of band (except those with $> \sim 80°$) have been used as guides for viscose. The rods were so mounted that the bands did not touch the fibers. These rods gave good results.

The orientation of the rod had much less effect if the conditions were less severe. We selected conditions such that these bands were not formed and the rods became quite smooth. These are the optimum conditions. They were attained by reducing the flame temperature and the rate at which the rod was drawn through the flame. The

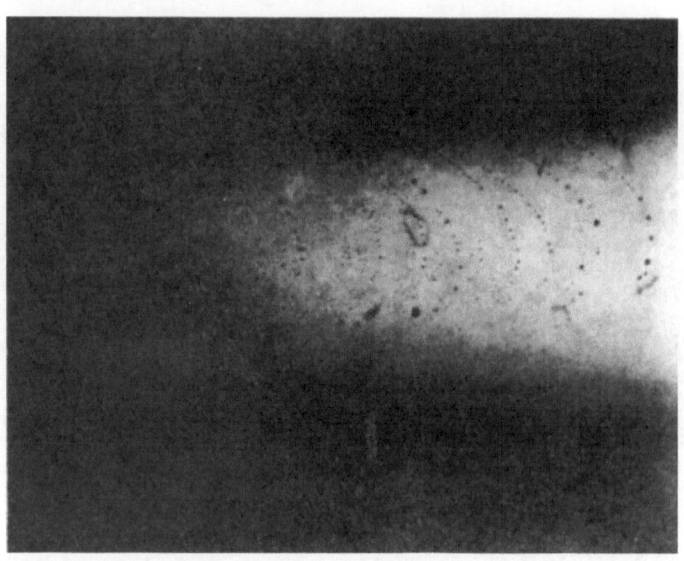

Fig. 52. Surface bands produced by flame polishing. × 30.

minimum temperature that produced polishing was found by inserting the rod until the end was level with the tip of the burner, with the temperature low. The oxygen and hydrogen flows were then gradually increased alternately until the tip began to melt. This temperature is a reference point for setting up the burner, which may then be increased appropriately to give a good polish at the preset withdrawal rate.

The polish was found to be best when the flame produced local overheating, such as to fuse only a thin surface layer. To produce this effect the working space in the muffle was made 40 mm in diameter, with the center of the flame at 75-80 mm. These conditions are suitable for routine polishing. Each unit has to be set up individually.

Sharp bands often appear on the surfaces if the conditions are nearly but not quite optimal. Figure 51 shows how these bands lie relative to the plane containing the optic and growth axes (this plane is denoted by the broken line). Each band is made up of a set of small ripples (Fig. 52), which run right along the rod.

Rods with these bands are entirely suitable for guides, because they have no sharp edges to damage the strands. To the unaided eye they appear ideally polished.

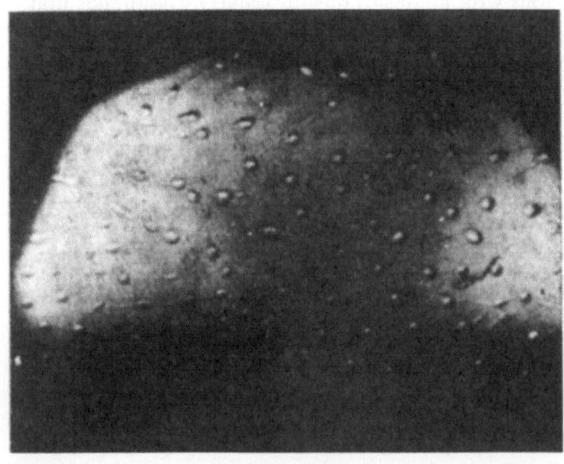

Fig. 53. Surface of a rod after a single polishing. × 70.

Polishing Rods of Various Diameters

The behavior of a rod depends on the diameter, if the conditions are severe; differences of only 0.1-0.2 mm in diameter require the gas flow or withdrawal rate to be changed. The larger the gas flow and the smaller the diameter, the greater must be the withdrawal rate, and conversely.

This relationship does not apply if the conditions are optimal. We have polished rods 4.0, 4.3, and 4.5 mm in diameter.

The surface finish was exactly the same if the polishing was repeated.

A single polishing never gives an entirely smooth surface; as a rule the surface is as shown in Fig. 53.

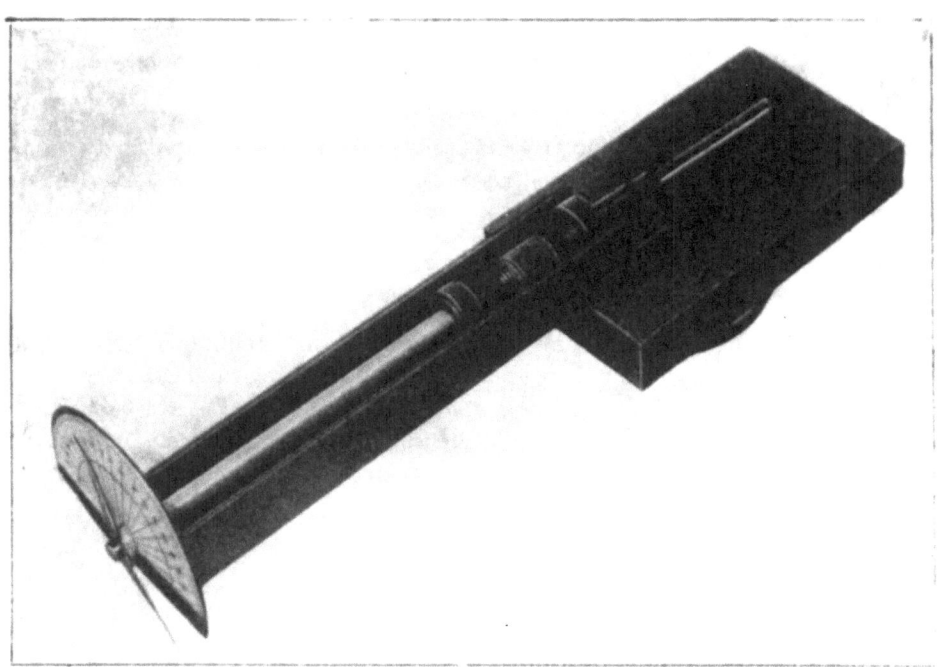

Fig. 54. Device for examining the surface polish.

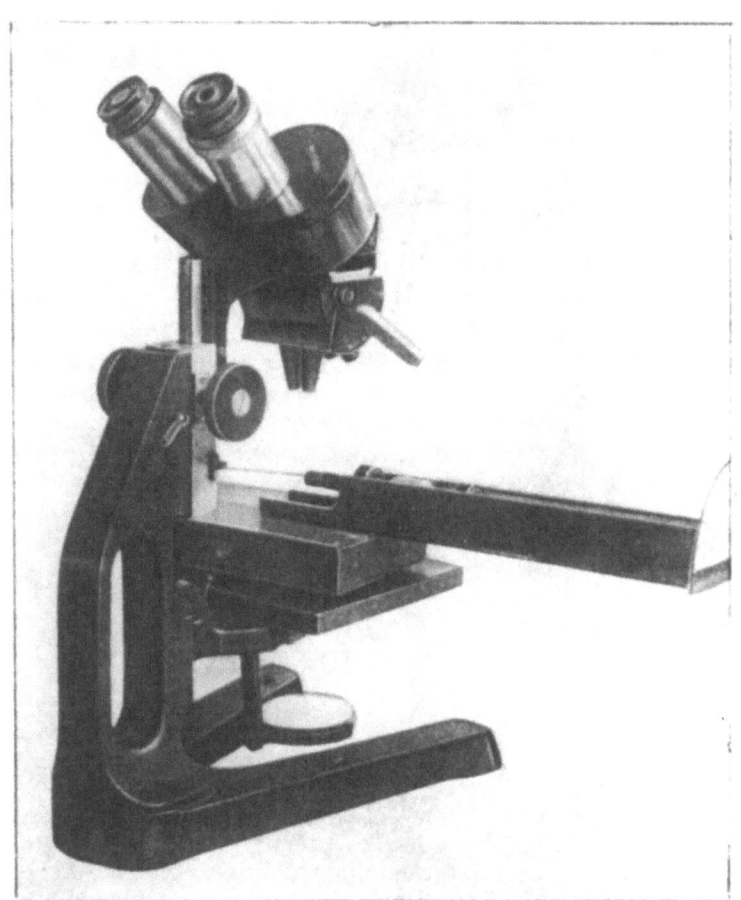

Fig. 55. A binocular microscope fitted with the device for examining the surface polish.

The best surface finish is obtained at the lowest temperature, when only the thin matt layer of adhering charge fuses, and if the rod is turned on its axis. The finish is improved as these conditions are approached.

Repeated polishing should be done not less than twice, without removing the rod from the holder. The finish should be made as good as possible by repeated polishing, if the guide is to be used with capron.

The rods to be polished must not be contaminated with abrasive dust (carborundum, fireclay, etc.), because the particles become stuck to the surface during the polishing and form hard, sharp projections. Rods with such projections are quite unsuitable for use as guides.

Each polished rod must be examined in reflected light to evaluate the surface finish. There are several instruments for use for this purpose with cylinders, but all are complicated.

We made a very simple device (Fig. 54) which was easy to use with rods of different lengths and diameters for viewing internal defects, for measuring angles between etch bands, and for examining surface polish. The rod was fitted into a rubber seating at the end of a metal tube. The tube turned on ball bearings set in slides which could move freely. The other end of the tube carried a needle which worked over a scale. The base was fixed to a circular table, for convenience in setting-up. The device was mounted on a stereoscopic viewer (Fig. 55).

Bending Corundum Fiber Guides

Corundum becomes plastic and can be bent at temperatures near its melting point. This property of corundum is not shared by other materials (e. g., agate), and is a very useful one, since curved guides are much used in making artificial fibers.

Fig. 56. The apparatus for bending rods.

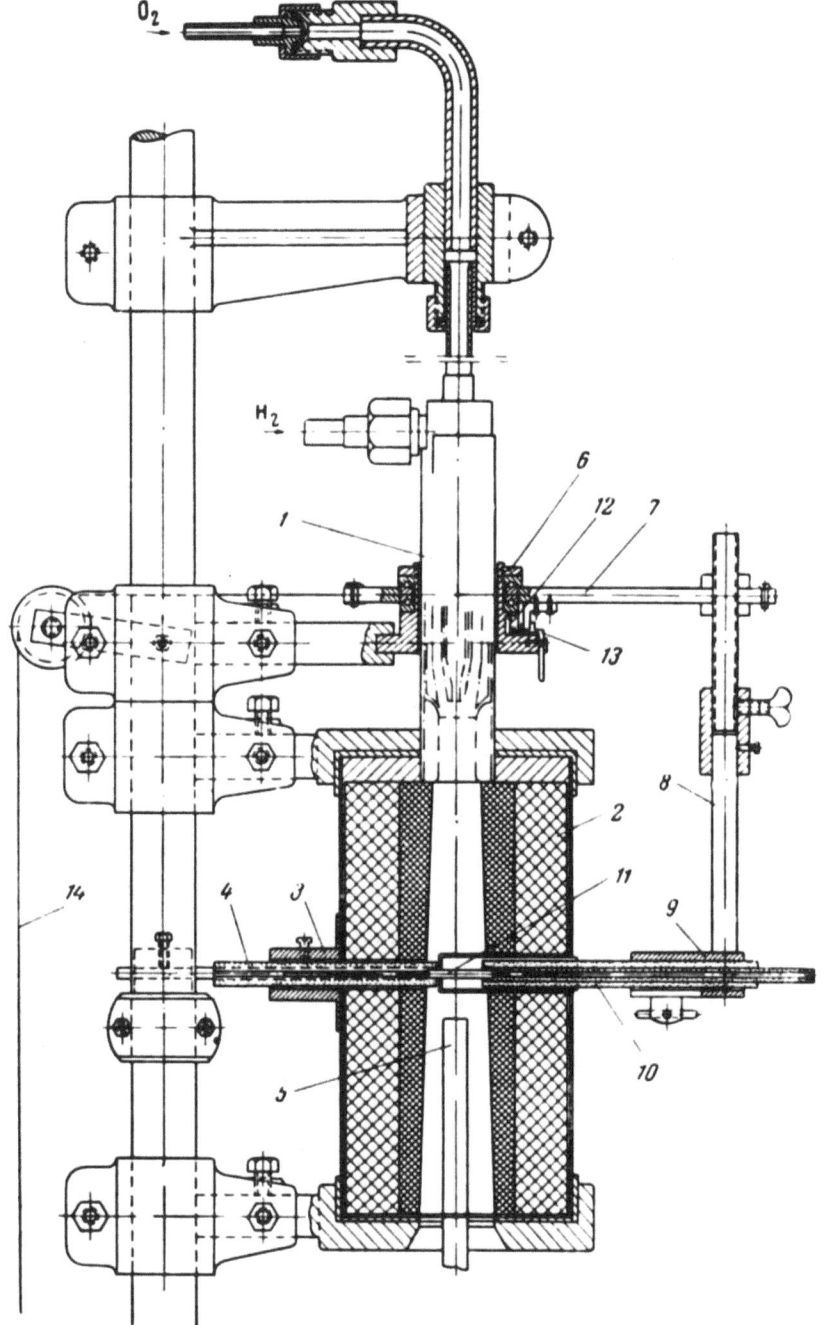

Fig. 57. The apparatus for bending rods: 1) burner; 2) muffle furnace; 3, 9) holders; 4, 10) refractory rods; 5) corundum candle; 6) bending mechanism; 7) moving disc; 8, 13) brace and lock; 11) corundum; 12) scale; 14) load wire.

In 1951-1952 we developed an apparatus (Figs. 56 and 57) for bending rods; this has the following main parts.

1. The oxyhydrogen burner 1. This is set in a muffle furnace and has a jet not smaller than 4.5-5 mm in diameter.

2. The muffle furnace 2, in which the hydrogen is burnt and the rods are bent. There are two holes in the muffle: one at the top for the burner, and a long, narrow horizontal one at the side, through which a refractory

147

holder is inserted with the crystal. The height of this slot is determined by the size of the refractory holder. A corundum candle 5 is inserted in the bottom of the furnace. There is a gap between the candle and the muffle sufficient to allow this part of the furnace to get hot. This candle is used to improve the cooling of the underside of the rod to be bent.

3. The mechanism 6, mounted in ball bearings, which has a movable disc 7 and brace 8, which supports the rotating holder, a guide-tube 9 to apply the load from the loading wire 14, a stop to restrict the deflection which is read on the scale 12, and a lock 13 for fixing the brace and tube 9 in a given position.

These devices are intended to reduce the careful and constant attention needed in bending rods to a given specification (bend angle, equal length of arms, orientation, etc.).

4. The refractory rods (10 and 4), which are held in holders 3 and 9 and which fasten the corundum rod 11. Silit rods are most convenient, because they are strong enough to support the loads applied, and are reasonably heat-resistant.

The holes in the silit rods must be round (never elliptical) or the rods will fracture during bending. The holes must have a clearance not greater than 0.1 mm. Too large a gap makes it difficult to clamp and remove the corundum. On this account the corundum rods are sorted into sizes for bending. Refractory rods with appropriate holes are made up for each size.

5. A device for withdrawing the rod. This is fixed to the left of the muffle at the level of the slot. The asbestos holder is kept very hot by the muffle, so the rod can be withdrawn as soon as it has cooled below a red heat.

6. A shutter for closing the slot during heating.

7. A valve block, for controlling and shutting down the flame.

8. Gauges to read the gas pressures in the lines, chambers and so on.

Pressure regulators and flow and pressure recorders are fitted to both gas lines in order to facilitate control and adjustment. The throttles for the hydrogen are 1.1-1.2 mm in diameter; those for the oxygen 16 and 1.3-1.5 mm in diameter. The flame and gas flows are set by trial for each unit.

A straight rod is bent by inserting the unpolished end in the moving refractory holder, which is locked in position. The rod is then heated in the flame from the window, and is slowly swung into the furnace; finally, it is inserted in the other (rear) refractory holder.

The desired angle of bend is set on the graduated scale fitted to the stop, while the rod is allowed to heat up for 1-2 min. The lock is released carefully. The rod starts to bend slowly under the load applied by the weight.

The moving holder reaches the stop when the rod has bent sufficiently. The holder is then released and the rod is extracted from the rear holder. The bent rod is brought slowly to the slot and allowed to cool; it is removed from the holder when it is no longer red.

There are some conditions which must be observed strictly in bending rods.

1. The furnace must be properly heated. Each time a rod is removed the shutter must be closed and 7-10 min allowed for the furnace to heat. A freshly started furnace must be allowed at least an hour to heat up.

2. The refractory holders must be coaxial; the slightest error will cause the rods to break.

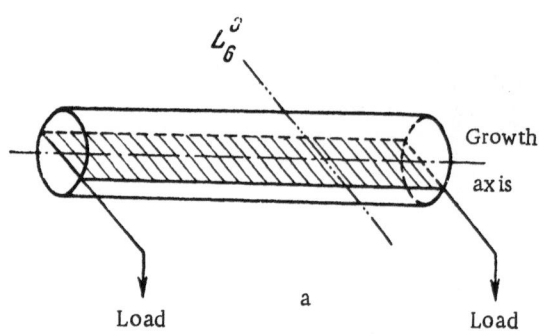

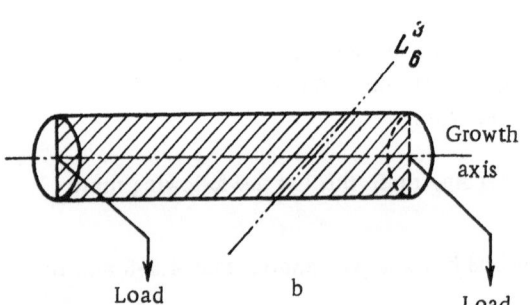

Fig. 58. Rod orientations used in bending: a) type I, b) type II.

3. The disc, brace, holder, and load wire must move freely and smoothly. Any sudden jerks will distort or break the rod.

Relation of Resistance to Bending to Crystallographic Orientation

The resistance to bending may be measured with a freely suspended weight fixed to the end of a rod heated nearly to its melting point.

There are two main ways in which the orientation and the action of the weight may be related:

I) the bending moment acts in the plane of the optic and growth axes (Fig. 58, a);

II) the bending moment acts in a plane normal to the plane of the optic and growth axes (Fig. 58, b).

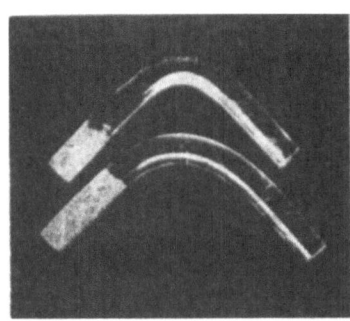

Fig. 59. Bends in two rods formed in planes at different angles to the plane of the growth and L_6^3 axes. × 1.

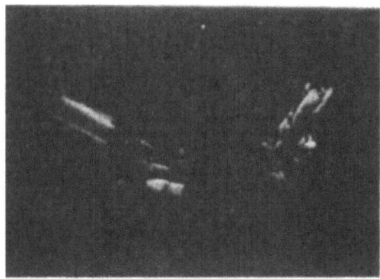

Fig. 60. Rod broken on bending in a plane normal to the plane containing the optic and growth axes.

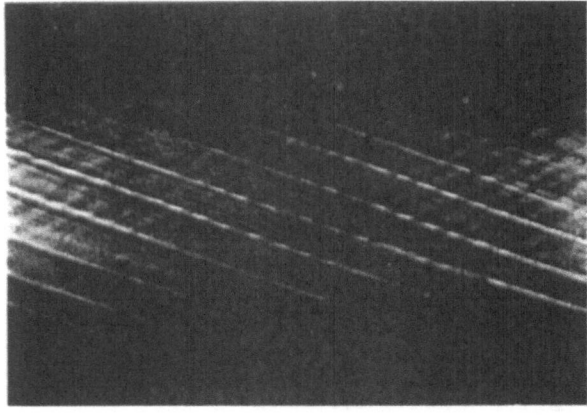

Fig. 61. Slip-plane traces on the surface of a bent rod, seen in reflected light.

The lowest temperature that can be used for bending is found by melting the tip of a vertical rod in the flame with the slot open.

The strength in bending was measured under two different conditions*

1) H_2 = 670 liters/hr, 2) H_2 = 800 liters/hr,

O_2 = 400 liters/hr; O_2 = 400 liters/hr.

The first provided local heating with the minimum general heating; the second provided more general heating.

The tests were done as follows. A rod 4 mm in diameter was put in the furnace and was heated for 1-2 min with the shutter closed, with flow rates H_2 = 700-800 liters/hr and O_2 = 400 liters/hr. The load used was any convenient one known to be less than the required value. This load was supplemented by adding 50 g balance-weights one at a time. It was found that a total load of 480-500 g bent the rod quickly, without danger of breakage. This load was used in all our bending tests.

A bend of 90° could be produced in 1-1.5 min with a rod of orientation I and with H_2 = 670 liters/hr, O_2 = = 400 liters/hr and a load of 480 g.

Rods of orientation II gave a different result. The 480 g load was too low, and 870 g was needed to bend the rods through the set angle, i. e., the load was nearly doubled. Even then the bending time was 10 min.

Rods of type II bend very strangely; at first they bend quickly, but after 30-32° the rate decreases, and the bending may stop altogether if the load is not increased.

At the higher temperature (H_2 = 800 liters/hr, O_2 = = 400 liters/hr) the type I rods bent in 40 sec with a 480 g load, i. e., much more quickly.

The type II rods also bent more quickly at the higher temperature, but the load still had to be larger. An increase of 200 g was needed.

* The temperatures were not measured. The H_2 and O_2 flow rates were used to give estimates.

Thus type I crystals show only about half the resistance to bending relative to type II crystals.

The following conclusions have been drawn from this work.

1. Bent rods of types I and II have very different radii of curvature.

Rods of type I give a radius of about 6 mm. Deviations are found only if the angle differs from 90°.

Rods of type II always give radii of 11.5-12 mm, i. e., values almost twice those for type I. The difference in the radii given by the two types is obvious to the unaided eye (Fig. 59).

2. Type I rods are more plastic and very rarely crack during bending, and then only because of malfunctioning of the apparatus or lack of care on the part of the operator. Rods of type II show so high a resistance to bending that it is rare for them not to crack. They often cleave as shown in Fig. 60 before the set angle has been reached. This is a very important feature to take account of in bending rods for guides. The losses by breakage can be minimized if the orientation is chosen properly.

3. Slip- plane traces are clearly visible when bent crystals are examined in reflected light (Fig. 61). This shows that corundum becomes plastic near its melting point, and that the plastic deformation occurs via slipping.

4. The resistance to bending depends on the diameter. The diameter of the rods used for the bending should for preference not exceed 4-4.2 mm. Fractures are rare and the resistance least if the crystallographic orientation is chosen properly.

Cutting Corundum Rods

A machine with a diamond saw can be used to cut the rods; so can one fitted with metal saws fed with abrasive (No. 100-120 carborundum). Our machine was fitted with a holder for the corundum rod (Fig. 62).

It is important to minimize vibration when the machine is working. There should also be no radial or axial play in the saw. Vibration and play can cause the rod to break. The machine should be checked to see that all the joints are tight, and that the crystal lifts smoothly. The ends of the rod should be cut and ground carefully, without jerks or knocks.

Fig. 62. Machine for cutting rods.

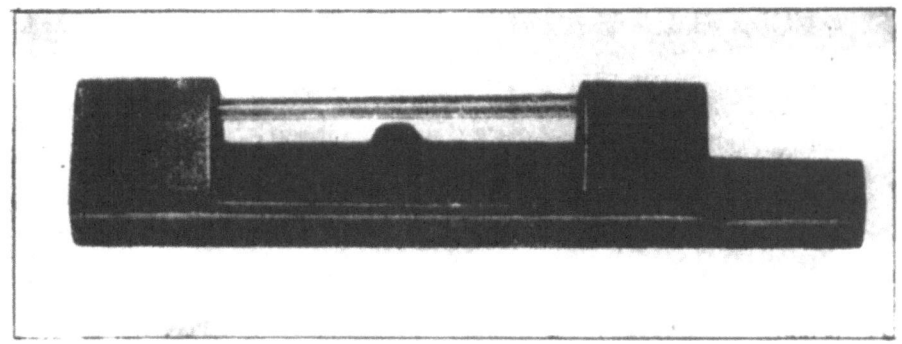

Fig. 63. Top fiber guide fitted in holder.

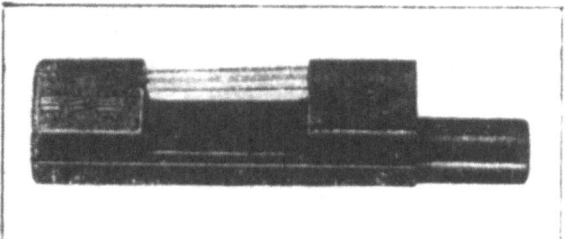

Fig. 64. Bottom fiber guide fitted in holder.

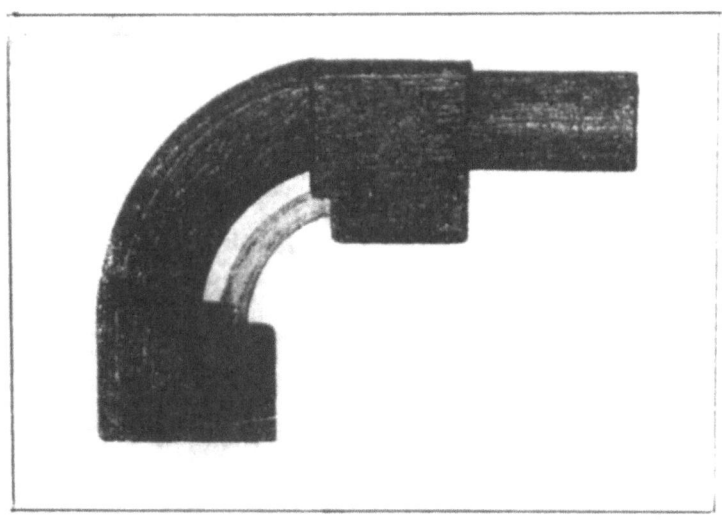

Fig. 65. Holder for bent fiber guide.

Some Types of Holder for Corundum Fiber Guides

The guides used for viscose are long and of small diameters, so special holders are required to fix them to the machine and to protect them from damage. Suitable holders have been designed and made.

Figures 63-65 show holders for various type of guide used in making artificial fibers.

1. The holder for the top guide set under the spinneret is shown in Fig. 63. The body of the holder contains a metal rod, because it has been found that the rod may break at the end where it is held to the machine, although the bending force (~ 3 kg) is small. The end of the metal rod is protected from the acid with lacquer or some other resistant compound. This type of holder has proved acceptable.

2. The lower guide is fitted in the holder shown in Fig. 64. Apart from dimensions, it differs in no way from the previous one. The body need not contain a metal rod, however, because it is shorter; the holder is therefore simpler to make.

3. The holder for bent control guides is shown in Fig. 65; the two ends must lie at an angle of 90°. This holder is fixed to the machine in the same way as either of the others.

4. The holder for straight control guides takes two rods, which are set at opposite ends at 180° to one another. We have designed several holders for these guides. They can be made of any plastic, because they are subject to only very small bending forces.

SUMMARY

The first monocrystal rod corundum guides were made in the Institute of Crystallography, Academy of Sciences of the USSR in 1949, and were sent for trial with various gauges of silk and capron. The guides proved satisfactory.

From 1953 onwards these guides were made in large numbers on the equipments described above, and were supplied in the holders described here. The corundum guides are being used more and more widely, and are displacing all the main types of guide made from other materials.

III. REVIEW AND DISCUSSION ARTICLES

DISLOCATIONS IN GERMANIUM MONOCRYSTALS

(REVIEW)

E. Yu. Kokorish

INTRODUCTION

Much attention has recently been given to growing highly perfect germanium monocrystals. Lattice defects in monocrystals reduce the carrier mobility, cause impurity atmosphere and space-charge effects, and increase trapping and recombination rates.

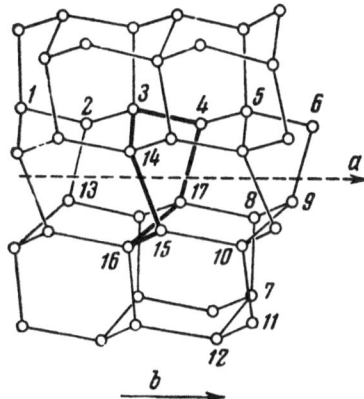 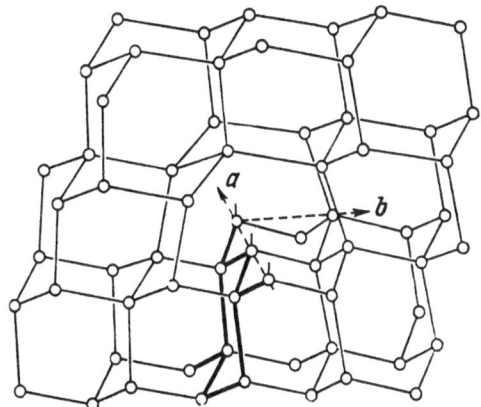

Fig. 1. A simple screw dislocation in a diamond lattice: a) direction of dislocation line; b) direction of Burgers vector.

Fig. 2. A simple 60° dislocation in a diamond lattice. The heavy lines indicate the extra half-plane: a) dislocation line; b) Burgers vector.

Line imperfections (dislocations) are one type of lattice defect. Dislocations have been discussed theoretically since 1930, but only recently has it become possible to observe them directly [1-3].

Read [4] and Cottrell [5] have dealt in detail with dislocation theory. We shall deal briefly with certain types of dislocation that may occur in diamond lattices, because germanium has this structure. Hoinstra [6] has shown that three types of simple dislocation are possible in diamond lattices, namely screw, edge and 60°. In all these cases the Burgers vector $\underline{b}$ equals half the length of a face diagonal, i. e., $^1/_2$[110], while the dislocation is directed along [110].

Figure 1 shows the approximate form of a screw dislocation in diamond. The (110) plane is the plane of the figure. The screw nature of the dislocation can be seen if we compare the normal hexagon 7-8-9-10-11-12-7 with the 13-2-3-14-15-16-17 structure. In the first case atom 7 is the start and end of a six-membered ring, whereas in the second there is a gap between 13 and 17. The distance from 13 to 17 is the length of the Burgers vector of the simple screw dislocation. The heavy lines 3-14-15-16-17-4-3 show one of the hexagons formed about the screw dislocation by distorting the lattice.

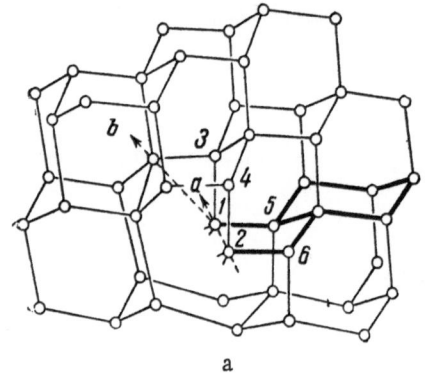
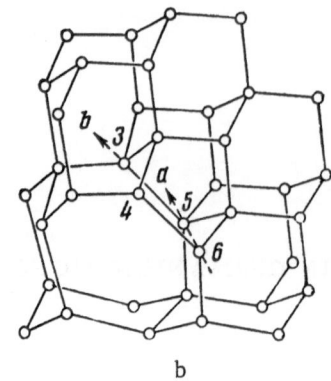

a

b

Fig. 3. Two simple forms of edge dislocation with (100) as glide plane. The thick lines show the extra half-plane: a) dislocation line, b) Burgers vector.

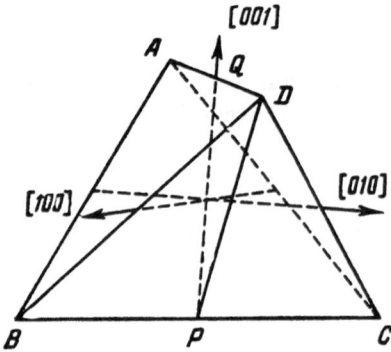

Fig. 4. Tetrahedron with its edges along [110]; AB = = [101], AC = [011], AD = [110], BC = [110], BD = = [011], and CD = [101].

Figure 2 shows a simple 60° dislocation. The vector $\vec{b}$ lies at 60° to the dislocation line. This type has a strong edge-component, but very little screw-component. The thick lines in Fig. 2 show the extra atomic half-plane, which stops within the crystal; some of the atoms are left with free valencies. Each such atom lacks one electron needed to complete a valence bond. These atoms may trap free electrons. The glide plane of this dislocation is (111).

Simple screw and 60° dislocations often result from plastic dislocation in germanium.

Figure 3 shows two simple forms of edge dislocation. The dislocation line has the [1$\bar{1}$0] direction, and $\vec{b}$ has the [110] direction. The glide plane is (001).

Figure 3a shows a type in which the extra atomic half-plane (thick lines) ends within the crystal and leaves each atom along the dislocation line with two unused valencies (see atoms 1 and 2).

Figure 3b shows another type, which has no atoms with free valencies along the dislocation line. This type can be produced from the first at high temperatures, when atoms 1 and 2 can migrate through the crystal. These atoms can fill vacant sites. The loss of atoms 1 and 2 caused paired valence bonds to form between atoms 3 and 4, and 5 and 6. The edge dislocation is then more stable, because there are no free valencies.

Other more complex dislocations can be built up from these simple types.

Consider the tetrahedron ABCD (Fig. 4), whose edge has the [110] direction. (The broken lines in Fig. 4 show the [100] directions.) We suppose that the Burgers vector $\vec{b}$ is directed along [$\bar{1}$10], i. e., lies in the BC edge. Then BC lies in the line of a simple screw dislocation. Edges DB, DC, AB, and AC lie in the lines of simple 60° dislocations, while AD lies in the line of a simple edge dislocation.

We can show that an edge dislocation in the direction of PD, [112], consists of two simple 60° dislocations CD, [101], and BD, [011]. We may write [101] + [011] = [112]. The edge dislocation PO, [001], consists of two simple 60° dislocations CA, [0$\bar{1}$1], and BD, [011], or BA, [$\bar{1}$01], and CD, [101], i. e., [0$\bar{1}$1] + [011] = [$\bar{1}$01] + [101] = = 2[001]. Table 1 gives the possible types of compound dislocation with Burgers vectors of $\frac{1}{2}$[110].

The simple dislocations (all three) are also given in the table. Hornstra has dealt in detail with compound dislocations, and with special dislocations in diamond lattices.

The boundaries between blocks that differ slightly in orientation are often detected in crystals. Bragg [6] and Burgers [7] in 1940 proposed a dislocation model for this effect.

Figure 5 shows such a crystal schematically. The slight difference in orientation causes the following effects at the interface:

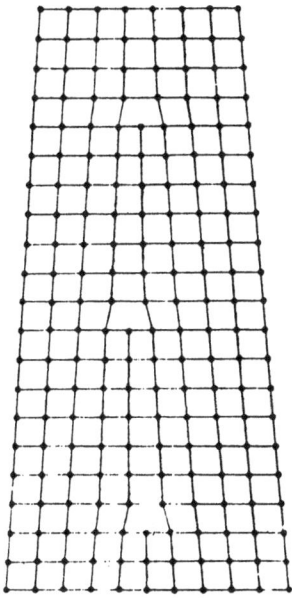

Fig. 5. Boundary between two blocks with a slight difference in orientation (after Bragg and Burgers).

1) the atoms at the boundary are displaced from their normal positions, and

2) there is present plastic deformation extending over several times the atomic spacing.

Some atomic planes (normal to the plane of the diagram) do not extend right through the crystal, but end at the boundary between the blocks and form edge dislocations. The atoms in such regions have a different environment. The distance $\underline{d}$ between dislocations along the boundary is given by

$$d = \frac{b}{\varphi},\qquad(1)$$

where $\underline{b}$ is the Burgers vector and φ is the relative rotation of the blocks (assumed small).

Vogel et al. [1] and others [8-10] have detected such boundaries.

The Dislocation Density

There are several methods of measuring dislocation densities. We shall consider some of them briefly.

Ideal crystals give extremely narrow diffraction lines [12]. Lattice defects cause extra x-ray scattering, and so broaden the lines. Double-crystal spectrometers are usually used to eliminate elastic broadening effects. One spectrometer crystal may be a perfect germanium crystal, and the other the specimen.

TABLE 1

Dislocations in a Diamond Lattice Whose Burgers Vectors are BC = $\frac{1}{2}$[110]

Dislocation line	Symbol	Angle between dislocation line and Burgers vector	Glide plane	No. of broken bonds per a*	
BC	[110]	0°	—	0	
AB, AC, DB, & DC	[110]	60°	(111)	1.41	
AD	[110]	90°	(100)	2.83 or 0	
BC + AC, BC + BA, BD + BC & DC + BC	[211]	30°	(111)	0.82	
AC + AB & DC + DB	[211]	90°	(111)	1.63	
AD + BD, DA + BA, AD + CD & DA + CA	[211]	73° 13′	(311)	2.45 or 0.82	
AB + DB & AC + DC	[211]	54° 44′	(110)	1.63	0
AC + DB & AB + DC	[100]	90°	(110)	2.0	0
AD + BC & AD + CB	[100]	45°	(110)	2.0	0
AC + BD & AB + C	[100]	45°	(110)	2.0	0

* $\underline{a}$ is the lattice constant, in cm.

If we assume the dislocations evenly distributed throughout the crystal, the dislocation density N_g is related to the angular broadening of the diffraction line by [13]

$$N_g = \frac{\beta^2}{9b^2},\qquad(2)$$

where $\underline{b}$ is the absolute value of the Burgers vector and β is the half-width of the diffraction line.

There are difficulties in determining the line-widths, and complicated x-ray equipment is needed.

A simpler method is [1] to count the etch pits. Kurtz et al. [14] have shown that the x-ray and etch-pit methods give the same result if there are less than 10^6 dislocations per cm^2. Selective etching gives underestimates at higher densities. Germanium monocrystals pulled from the melt have densities lower than 10^6 cm^{-2}, and so etching methods are suitable.

TABLE 2

The Best Etching Reagents for Revealing Dislocations in Germanium

No.	Composition	Uses	Etching time
1	2 cc HNO_3 + 4 cc HF + 4 cc H_2O + 200 mg $Cu(NO_3)_2$	Chemical polishing and etching on (100)	6-7 min, in the cold
2	10 cc HF + 10 cc H_2O_2 + 40 cc H_2O	Detecting dislocations on (111) and (110)	3-6 min, in the cold
3	12 g KOH + 100 cc H_2O + 8 g $K_3[Fe(CN)_6]$	The same	2-6 min, boiling
4	50 cc HNO_3 + 30 cc CH_3COOH + 30 cc HF + 0.6 cc Br	Chemical polishing and etching on (111)	

The atomic planes are distorted around dislocations; in some places the bonds are lengthened, in others shortened. The energy is the sum of the elastic deformation energy stored around the dislocation and the energy of the broken bonds at the edge of the dislocation. This energy excess makes the crystal less stable near a dislocation, and so more liable to etching [15]. Hence, a carefully polished and appropriately etched surface will give etch pits at the points where the dislocations emerge. The shape of the pits depends on the crystallographic direction and on the etching agent.

Figure 6 shows typical etch-pit shapes for various crystallographic directions [16].

Ellis [2, 16, 17] and Vogel [1, 18] describe in detail etch methods of detecting dislocations.

The following points must be borne in mind in etch studies.

1. Dislocations lines seen on (111) may not penetrate far into the crystal. They may form a three-dimensional net of dislocations, which net splits the separate blocks up into micromosaics.

2. Etch pits seen on (111) and (100) planes may not correspond to the same type of dislocation, because little is yet known about the effects of etching on different types of dislocation.

3. The etching agent must be selective; it must act when the dislocation line meets one type of plane, but not when it meets another type [2].

4. The surface must not deviate from true (111) or (100) planes by more than a few degrees. A small deviation from (111), for instance, causes the pits to become elongated (instead of equilateral), and so the effective density is reduced. The etch pits cannot be seen at all if the deviation from (111) is more than 15°.

There are methods of detecting dislocations other than those dealt with above. For silicon Dasch [19] has proposed to plate the section with copper, followed by observation in transmitted infrared light.

Recently Tyler and Dasch [20] have detected dislocations in deformed germanium in the following way. Lithium is deposited on the surface. The sample is then heated at 400-650°C for 100 hours. The lithium diffuses into the crystal mainly along dislocation lines. The specimen is then polished and is etched for 20-60 sec (etch in 4 parts HNO_3, 2 parts HF and 15 parts CH_3COOH). The etching occurs mainly at the points where the lithium is concentrated (i. e., at the dislocations).

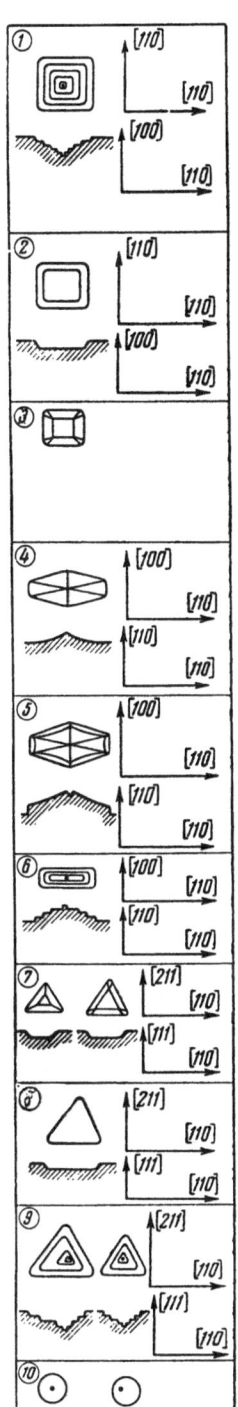

Fig. 6. Typical etch-pit shapes for various crystallographic directions. Pits Nos. 1, 2, and 3 are produced on (100) by etch No. 1 (Tables 1 and 2). Type 1 pits. A typical feature is that there is a central hole, which lies at the point where the dislocation emerges. Irregularities may reveal unevenness in the dislocation. The pits are about 25 μ along an edge. Type 2 pits. A typical feature is that there is no central hole. The sizes vary greatly (~ 2-20μ). Type 3 pits are still smaller. A typical feature is that there is a raised center. Pits types 4 and 5. These probably indicate places where dislocation lines meet the surface. Types 4 and 5 have sizes of ~ 10 × 20 μ. Type 6 pits are small, and are probably early forms of type 5. Pits types 7, 8, and 9 are produced on (111) by etch No. 2 or 3. Type 7 has no central holes. The sizes vary up to 15 μ along an edge. Type 8 pits are probably the final forms of type 7. They are up to 40 μ in size. Type 9 pits are seen from the ellipses of edge dislocations. The steps are caused by unevenness in the dislocations. The middle point is not always at the center of the outer steps. Type 10 pits are produced on (111) by etch No. 4, and are caused by edge dislocations.

The Origin of Edge and 60° Dislocation During Growth

Dislocations in germanium can occur in the following ways:

1) during slip caused by stresses; this can occur while the crystal is cooling to room temperature, or, especially, to 500°C, because germanium is plastic above 500°C [21, 22];

2) when vacancies accumulate during growth [23, 24];

3) when dissolved atoms are unevenly distributed, and occur locally in concentrations high enough to stress the lattice severely [25].

Germanium monocrystals grown by pulling from the melt are usually used to make semiconductor devices [26]. Germanium expands as it solidifies; if a crucible is used, its walls exert a pressure on the germanium. The crystal is deformed as it cools through its plastic range, and many dislocations are formed. Such crystals cannot be used to make semiconductor devices.

Crystals grown by pulling are not confined in crucibles, but still have defects. The neck (the part where the crystal expands from the seed to full size) is often twinned, for instance. Some of the latent heat travels along the axis to the seed (the cold end), and some is lost laterally. There are axial and radial temperature gradients, which may cause high stresses.

The stress F is related to the temperature gradient by

$$F = E\alpha T \frac{\partial T}{\partial r}, \tag{3}$$

where E is Young's modulus, α is the linear expansion coefficient, and T is the temperature.

These stresses may exceed the yield stress (often about 1 kg/mm² for monocrystalline metals) if the temperature gradient is large.

The dislocation density N_g is related to the temperature gradient by [21]

$$N_g = \frac{\alpha}{b} \cdot \frac{\partial T}{\partial r}, \tag{4}$$

where $\underline{b}$ is the Burgers vector and $\partial T / \partial r$ is the temperature gradient (radial or axial).

159

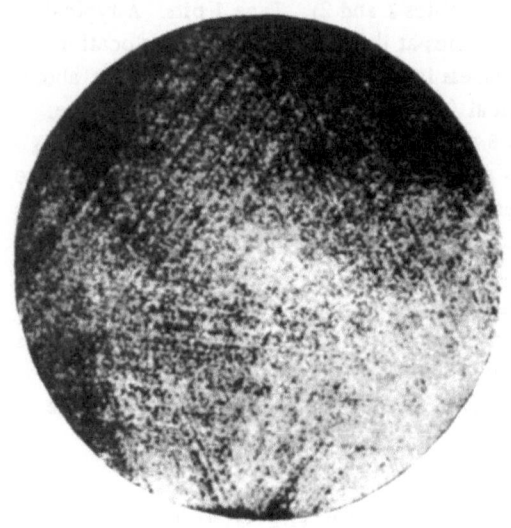

Fig. 7. Concentration of etch pits along [110] directions. × 8.

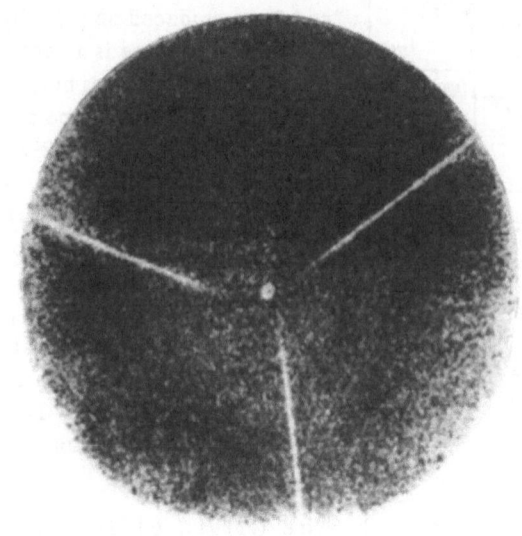

Fig. 8. Ray-like distribution of etch pits along three directions. × 8.

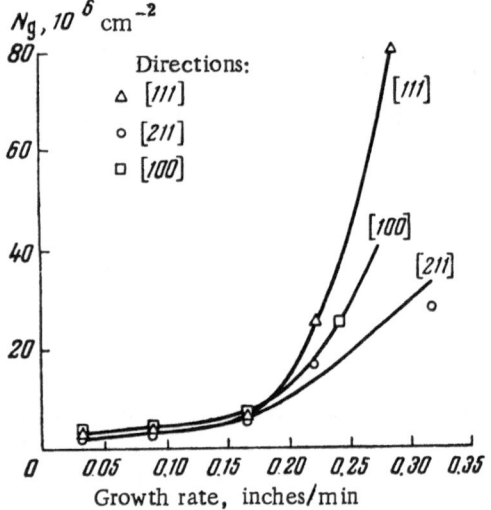

Fig. 9. Dislocation density vs. growth rate for three crystallographic directions:

Now, (4) is a simplified formula, because the real density has a tensor form. Indenbom [27] has dealt with this topic in detail.

Billig [28], Dorendorf [29], Bennet and Sawyer [30] and Cressel and Powell [31] have studied how dislocations form in germanium monocrystals. Billig has studied how the dislocation density in a monocrystal varies with the temperature gradient. He has found that the two increase together. He has also found a three-rayed system of etch pits on the [110] directions (Fig. 7).

Dorendorf has also found a ray-like distribution of etch pits along the three [211] directions (Fig. 8). The middle of a crystal usually shows a random distribution.

Billig and Dorendorf explain this type of pattern by saying that dislocations cannot arise at the crystallization front, because all parts of the front experience the same conditions (temperature, impurity concentration). They suppose that the dislocations arise as the crystal cools to room temperature, because different parts cool at different rates. The tendency of growth pits to concentrate along certain directions is related to the growth conditions. The isotherms are convex when the dislocations lie in the [110] directions, for instance. The radial temperature gradient in the crystal is given by (4).

In any given cross section the periphery is cooler than the center. The periphery therefore contracts more than the center, and compresses it. Now the center is the hotter, and so may be deformed. Dislocations thus occur along the [110] directions. The glide planes are (111) planes.

The ray-like distribution along the [211] directions occurs when the solid-liquid interface is concave towards the solid. This occurs if the crystal is increasing in diameter. The center is again under pressure. But here the center is the cooler, and so puts up much more resistance than in the previous case. Then it is the periphery that is deformed, and the dislocations appear in rays along the [211] directions. These rays occur mostly at the top of the crystal. They penetrate farther and farther within the crystal as the diameter increases. One or two rays only are sometimes seen.

The dislocations show a random distribution in the middle of a crystal, and there is usually no great difference in N_g between the center and the edge. The isotherms are nearly flat, and the dislocations are caused by the axial gradient alone.

Kurtz et al. [32] have studied the relation of dislocation density to growth rate for germanium monocrystals grown by zone fusion. The density and growth rate increased together. Figure 9 shows the dislocation density vs. growth rate for three crystallographic directions. Above 4 mm/min the density rises rapidly with the growth rate.

Kurtz et al. assume that here the atoms do not have time to take up their proper positions in the lattice when the growth rates are large. There are then also many vacancies, which can give rise to dislocation loops, as Chalmers [33] supposes.

The large temperature gradient and rapid cooling associated with rapid motion of the front may cause large stresses, which in turn increase the dislocation density. Thus, increase in growth rate increases the dislocation density.

Conditions such that no large temperature gradients occur during cooling must be ensured if perfect germanium crystals are to be grown with low dislocation densities.

The Effects of Dislocations on the Electrical Behavior of a Crystal

Gallagher [21] and Pearson [34] have shown that the edge dislocations produced by plastic deformation cause acceptor levels in the forbidden band, which levels act as recombination centers. Figure 10 shows a 60° dislocation which produces a line of unused valence bonds (one unused valence electron per atom). Shockley call these "dangling" bonds [35]. Read [36-38] has given a detailed theory of these bonds as acceptor centers.

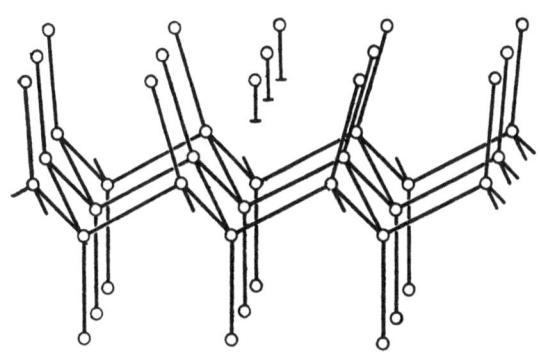

Fig. 10. Unused valence bonds (afterShockley).

A conduction electron may be trapped by an unused bond. Its energy E_2 will still be greater than that of an electron in the valence band, but will be less than that of a conduction electron. Then E_2 is such as to lie within the forbidden band. In n-type material, the Fermi level E_f lies above E_2. Edge and 60° dislocations produced in any way (e. g., by plastic deformation) cause acceptor levels, which trap conduction electrons. These electrons cause the dislocations to become negatively charged lines. The acceptor centers lie close together along these lines, but the E_2 levels are separated by distances equal to the lattice constant of germanium. The degree of filling is given by a function f:

$$f = \frac{1}{1 + \exp \dfrac{E_2 - E_f}{kT}}, \tag{5}$$

where k is Boltzmann's constant and T is the absolute temperature.

The unused bonds become of covalent type when they have trapped electrons. Let a be the distance between two such bonds. If a is small relative to the mean distance between donor impurity atoms, the dislocation acts as a line charge of strength q/a per unit length.

This negative line charge repels the conduction electrons. A cylinder of positive space charge results.

Little scattering occurs for electrons moving parallel to the dislocation line.

The scattering is larger if the motion is normal to the line; the drift tracks are strongly curved near dislocations.

Dislocations produce stress fields, which may interact with other stress fields in the crystal. The dislocations may also tend to take up impurity atoms, and so reduce the general stresses. Cottrell [5] has shown that the energy of interaction between a dislocation and an impurity atom is given approximately by

$$U(r\varphi) = -\Delta V \frac{\mu b}{3\pi} \cdot \frac{1+\nu}{1-\nu} \cdot \frac{\sin \varphi}{r} \tag{6}$$

where ΔV is the volume change caused by the impurity atom, μ is the shear modulus, ν is Poisson's ratio, $\underline{b}$ is the Burgers vector, φ is the angle between this vector and the line joining the impurity atom to the center of the dislocation, which latter line is of length $\underline{r}$.

Impurity atoms whose radii are larger than that of germanium are drawn to the lower part of a positive dislocation, and those that are smaller to the upper part. Figure 11 shows the possible positions of impurity atoms near an edge dislocation. In case $\underline{a}$, the dislocation is free from impurity atoms. In case $\underline{b}$, boron gives the impurity atmosphere, in case $\underline{c}$, antimony.

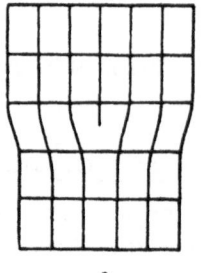 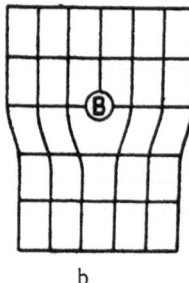 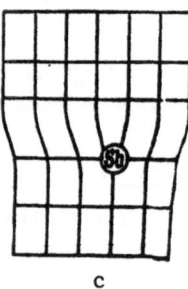

a   b   c

Fig. 11. Possible arrays of impurity atoms near a dislocation.

TABLE 3

Impurity	Atomic radius	U, ev
Cu	1.28	0.12
As	1.16	0.12
Al	1.43	0.36
Sb	1.34	0.20
Sn	1.45	0.48
P	0.93	0.54
Si	1.18	0.06

Table 3 [39] gives interaction energies for edge dislocations for certain impurities with r = 4 A.

It is considered that dislocations formed during cooling have more impurities around them than do ones produced by plastic deformation [40]. The cause is that the diffusion rate is higher near the melting point than it is at the temperatures used in plastic deformation experiments. The impurity concentrations near a dislocation reduce the effectiveness of the dislocation as a recombination center.

The impurity atoms may surmount the potential barrier U that holds them to the dislocation, and may then move through the the crystal. The probability of such escape is proportional to $\exp \frac{-U}{kT}$. Now if kT > U, the atoms can move freely about the crystal. The unused bonds in the dislocations then become accessible again, and so trap conduction electrons. The dislocation becomes a negatively charged line surrounded by a positive space charge. The dislocation acquires a higher recombination cross section and so the minority-carrier lifetime is reduced. The impurity atmosphere diminishes this effect. The dislocation also affects the degree of ionization of these impurity atoms. Thus, Cu atoms form a neutral atmosphere about a dislocation; at high temperatures they move freely about the crystal.

If the crystal is cooled rapidly from a high temperature, these randomly distributed atoms are trapped, and cannot return to the dislocations. They can return only if the cooling is slow. This indicates one cause of the changes in carrier lifetime produced by heat-treatment.

If we assume that each dislocation acts as a recombination center, the mean lifetime τ is related [41] to the dislocation density by

$$\sigma = \frac{1}{\sigma_R N_D}, \tag{7}$$

162

where σ_R is the efficiency of recombination per unit length of dislocation. Low-resistant germanium has $\sigma_R = 35 \cdot 10^{-3} \, cm^{-1} \cdot sec^{-1}$, and high-resistant material has $\sigma_R = 5 \cdot 10^{-4} \, cm^{-1} \cdot sec^{-1}$.

Pearson et al. [34] have shown that edge dislocations cause acceptor levels 0.2 ev below the conduction band. J. McKelvey [41] has shown that the cylindrical recombination regions are 2.8 A in diameter for electrons in p-type germanium and 1.15 A for holes in n-type germanium, for crystals grown in the normal way from the melt. Morison [42] has dealt in detail with electron-hole combination at dislocations. He has proposed a model to explain the experimental data. He has also shown that the space charge near a dislocation affects the recombination very much. In particular, it causes a slow fall in the photoconductivity, as is found in silicon and in n-type germanium.

Kurtz et al. [14] have related the lifetime to the dislocation density for germanium monocrystals.

Wertheim and Pearson [43] have related the lifetime to the dislocation density for germanium n- and p-type monocrystals that had been plastically deformed. The p-type material at room temperature followed the law

$$\tau = 0.7 / N_D. \tag{8}$$

The n-type (for holes) gave the relation

$$\tau = 2.5 / N_D. \tag{9}$$

These results show that the capture radius for electrons is some two or three times larger than is that for holes.

Removal of Dislocations by Annealing

We now consider some aspects of reducing the dislocation density by high-temperature annealing. The dislocations become more mobile as the temperature rises. They can move about the crystal. Read [4] has considered in detail how dislocations move and interact. They repel one another or recombine, depending on whether they are of the same or different signs.

Dorendorf [29] has annealed pulled crystals in vacuo at 800°C for 40 min. The dislocation density was then measured by etching methods. He found that the density was reduced by 60% by the treatment.

Greiner et al. [44] have annealed slightly deformed germanium specimens. The temperature was 725°C, the time 1 hour. The density was roughly halved.

Little has yet been done on the reduction in dislocation density caused by annealing. Several hours at 700-800°C is clearly not sufficient, because the dislocations do not vanish entirely.

Present efforts are directed to removing the causes of the dislocations that form during growth. To this end sudden changes in temperature and growth rate are avoided, and the growth conditions are made such that the axial and radial temperature gradients do not cause thermal stresses sufficient to exceed the elastic limit. The boundary between the solid and liquid phases must be flat in order to ensure this [45]. The seeds must also be perfect monocrystals.

It is stated [30, 31] that, if these conditions are complied with, it is possible to make germanium monocrystals with dislocation densities below 100 cm^{-2} by zone fusion.

I wish to thank Dr. Sheftal' for looking over the manuscript, and for discussing some topics.

LITERATURE CITED

[1] F. Vogel, W. Pfann, H. Corey, and E. Thomas, Phys. Rev. 90, 489 (1953).

[2] S. G. Ellis, J. Appl. Phys. 26, 1140 (1955).

[3] A. J. Forty, Direct Observation of Dislocations in Crystals [Russian translation] (Moscow, 1956).

[4] W. T. Read, Dislocations in Crystals (McGraw-Hill Co., N. Y., 1953).

[5] A. H. Cottrell, Dislocations and Plastic Flow in Crystals [Russian translation] (Moscow, 1958).

[6] J. Hornstra, J. Phys. Chem. Solids 5, 129 (1958).

[7] W. Bragg, Proc. Phys. Soc. 52, 4 (1940).

[8] J. Burgers, Proc. Phys. Soc. 52, 23 (1940).

[9] J. Okada, J. Phys. Soc. Japan 10, 1018 (1955).

[10] R. Hashiguchi and E. Matsuura, J. Phys. Soc. Japan 12, 1347 (1957).

[11] W. Pfann and L. Lovell, Acta Met. 3, 512 (1955).

[12] R. W. James, Optical Principles of the Diffraction of X-rays (London, 1950).

[13] S. Kulin and A. Kurtz, Acta Met. 2, 354 (1956).

[14] A. Kurtz, S. Kulin, and B. Averbach, Phys. Rev. 101, 1285 (1956).

[15] J. Allen, J. Electronics 23, 439 (1956).

[16] S. Ellis, Coll.: Transistors. I (RCA Laboratories, 1956), p. 97.

[17] S. Ellis, J. Appl. Phys. 28, 1262 (1957).

[18] F. Vogel, Acta Met. 5, 377 (1957).

[19] W. Dasch, J. Appl. Phys. 27, 1193 (1956).

[20] W. Tyler and W. Dasch, J. Appl. Phys. 28, 1121 (1957).

[21] G. Gallagher, Phys. Rev. 88, 721 (1952).

[22] R. Treuting, Trans. AIME 203, 1027 (1955).

[23] E. Teghtsoonian and B. Chalmers, Canad. J. Phys. 29, 370 (1951).

[24] E. Teghtsoonian and B. Chalmers, Canad. J. Phys. 30, 388 (1952).

[25] R. Newman and W. Tyler, Phys. Rev. 96, 882 (1954).

[26] J. Czochralski, J. Phys. Chem. 93, 219 (1918).

[27] V. L. Indenbom, Kristallografiya 2, 594 (1957).

[28] E. Billig, Proc. Roy. Soc. 235, 37 (1956).

[29] H. Dorendorf, Z. angew. Phys. 9, 413 (1957).

[30] A. Bennett and B. Sawyer, Bell. Syst. Tech. J. 35, 637 (1956).

[31] I. Cressell and J. Powell, Progress in Semiconductors. II (London, 1956), pp. 137-164.

[32] A. Kurtz, S. Kulin, and B. Averbach, J. Appl. Phys. 27, 1287 (1956).

[33] B. Chalmers, Amer. Inst. Mining Met. Engr. 200, 519 (1954).

[34] G. Pearson, W. Read, and F. Morin, Phys. Rev. 93, 666 (1954).

[35] W. Shockley, Phys. Rev. 91, 228 (1953).

[36] W. Read, Phil. Mag. 775 (1954).

[37] W. Read, Phil. Mag. 45, 1119 (1954).

[38] W. Read, Phil. Mag. 46, 111 (1955).

[39] A. Kurtz and S. Kulin, Acta Met. 2, 352 (1954).

[40] J. Allen, J. Electronics 1, 580 (1956).

[41] J. McKelvey, Phys. Rev. 106, 910 (1957).

[42] S. Morrison, Phys. Rev. 104, 619 (1956).

[43] G. Wertheim and G. Pearson, Phys. Rev. 107, 694 (1957).

[44] E. Greiner, P. Breidt, J. Hobstelter, and W. Ellis, J. Metals 9, 813 (1957).

[45] N. N. Sheftal', N. P. Kokorish, and S. Kh. Mukhonkin, Paper at the First Conference on Crystal Growth [in Russian] (Moscow, March 5-10, 1956).

THE MAIN TRENDS IN THE STUDY OF INORGANIC CRYSTALS CONTAINING ORGANIC IMPURITIES

(REVIEW)

E. N. Slavnova

Minute traces of certain organic compounds can influence the course of crystallization, or can produce new properties in the growing crystals [1]. Although the effects are of great practical and theoretical interest, little has been done on the interactions of organic compounds with the faces of growing crystals.

There is, for instance, no general agreement on whether the compounds must enter the crystal to produce their effects, or whether their influence on the surfaces is sufficient. It is also uncertain whether the organic compound is present in the host crystal in molecular or in crystalline form, and what the type of the mixed system may be. Do solid solutions form, or do we get anomalous mixed crystals, or do we get an altogether new type of system? The state of the subject is unsatisfactory because it is most hard to study the complicated mechanisms of the processes.

The technical difficulties caused by the smallness of the concentrations, whether in the solution or in the crystal, are the reason why the states of such compounds in the crystals are discussed in terms of theoretical ideas and of indirect evidence, rather than in terms of direct experimental data. This reason, and the purely intellectual interest of the interactions between traces of organic compounds and inorganic crystals, are the causes of the many trends shown in this field, which overlaps many specialities.

The extensive topical literature is extremely scattered. It lacks a common purpose, and even a common terminology, so it is very difficult to survey. Therefore we have tried to outline the main trends which have influenced current views on the subject. We have given special attention to papers that deal with the states of impurities in crystals. The topic is important because these states result from the interactions between the components, and so can give us a better understanding of the molecular processes at the surfaces of growing crystals. They also may explain some of the new properties the crystals show.

We have surveyed the following trends:

1) morphologic, including studies on dyes in crystals and on epitaxis;

2) crystal genesis (certain aspects);

3) structure and formation mechanism studies for anomalous mixed systems; and

4) work related to the adsorption of dyes on crystal faces.

We deal also with the trend begun by Vedeneeva, on the spectrophotometry of dyes adsorbed on clay minerals [2].

1. Trends Related to Morphologic Studies on Crystals Colored by Dyes

The earliest studies on organic impurities in inorganic crystals were concerned with morphology. Organic dyes were used to study the effects on the properties, mainly for crystals of low symmetry; the dyes were readily detected visually.

Senarmont [3] was one of the first to show that dyes were taken up by growing crystals. He used strontium nitrate (monoclinic), with saffranin, logwood and other organic substances. The crystals showed considerable pleochroism.

Gaubert did much work on this topic. At first [4, 5] he dealt with optical anomalies in the cubic crystals of barium and lead nitrates, which he grew from solutions containing methylene blue. He found a strong blue-to-violet pleochroism with barium nitrate. This effect he explained in terms of oriented microscopic dye crystals within the barium nitrate. The two extreme colors correspond to two directions in the indicatrix; the violet appears when vibrations parallel to N_g (γ) are absorbed, and the blue when those parallel to N_p (α) are absorbed.

These mixed crystals (these syncrystallizations) were explained as arising from the close similarity between the structure parameters. Later Gaubert [6, 7] expressed the view that syncrystallization was responsible for the changes in habit caused by impurities. The submicroscopic dye crystals formed on certain faces retarded the growth, and so these faces began to predominate. Lead and barium nitrates grow as cubes when methylene blue is present, and the dye is deposited on the cube faces. But the dyed crystals did not always show the pleochroism proper to the dye, and so Gaubert concluded that the dye exerted its effect at the molecular level.

The dye may exist in two distinct states in the host crystal [8], namely:

1) in a state the same as that in the solution (solid solution), and

2) in a crystalline state, as an intergrowth. The first state is found when the dye concentration is low; the second occurs only near the saturation point.

Bunn [9] and Royer developed Gaubert's ideas on the importance of structural similarity to syncrystallization.

Bunn replaced Gaubert's view that high concentrations were needed to cause intergrowth by a theory of oriented adsorption. This adsorption can cause regular intergrowths even at low concentrations. The main requirement is that the planes of the two substances should be very similar.

Bunn sought to relate oriented deposition, adsorption and habit changes. He showed that rock salt could form oriented intergrowth with urea, $KClO_3$, with $KMnO_4$, and lead nitrate, with methylene blue. Later Bunn [10] expressed the view that mixed crystals are responsible for habit changes only at high dye concentrations. The effects occur at the molecular level at low concentrations. The large dye molecules attach themselves via their polar groups to the surface, and, because they are large, prevent material from reaching the surface, and so slow down the growth.

Bunn's theory, given in its most highly developed form by Buckley [11], amounts to the following concepts: impurity ions, if present at low concentrations in the solution, can form stable two-dimensional complexes on certain faces, which complexes become unstable three-dimensional ones as growth proceeds. These latter break up rapidly, and growth continues only slowly. On other faces, however, stable compounds may form, and so the dye is incorporated, with less effect on the growth rate. At low concentrations the dye interacts with one kind of face (because it is present as ions), whereas at high concentrations it may affect other faces.

Buckley deals with 600 dyes used with some inorganic crystals and remarks that, although Bunn's theory cannot explain many details of habit changes, it is the best theory available. "It is probably nearer to the truth than any theory yet proposed" (Crystal Growth, p. 273, Russian edition). Even Buckley [12, 13] doubts whether a dye can form a crystalline intergrowth with an inorganic substance, however, because it is difficult to see how large dye molecules can resemble the structure elements of ionic crystals, i. e., it is difficult to believe that the anion-to-cation distances in the dye and in the host crystal can be similar. Moreover, it is found that the dyes are taken up best from solutions of low concentration (10^{-3} M). "It is improbable that the dye particles could deposit side-by-side and form independent small crystals within the main crystal at such low concentrations" (Crystal Growth, p. 259, Russian edition).

Buckley supposes that most of the habit changes caused by dyes occur because the dye is present in a molecular or ionic state. The dye dissociates into simple and complex ions. The latter are adsorbed on certain faces, and are oriented; they are on the average distributed uniformly, but are far apart. They are oriented in the same way because they adhere to the surface in the same way. Buckley considers that the large dye molecules (ions) are held to the surface only by their active groups, e. g., SO_3Na, COONa, COOH, etc. These groups must have their oxygen atoms disposed in nearly the same way as do the anions of the host. With sulfonic acid dyes, for instance, the SO_3 group can replace ClO_3 in $NaClO_3$ crystals. In such cases these groups are held by oxygen triangles. More often, however, they are held by the oxygen pairs on certain edges in the SO_3Na polyhedra. According to Buckley (and Bunn), the dye affects the habit because it disturbs the supply of material to the growing faces.

But Buckley observes [14] that the dye sometimes speeds up the growth of faces on which it is deposited. This is true for lead nitrate and methylene blue. Thus Buckley concludes that the pleochroism often seen in dyed crystals is caused by oriented arrays of dye molecules. Buckley's numerous studies led him to conclude that it is difficult to classify dyes in accordance with their effects on the habit, and that much still remains inexplicable. Chemically related dyes have quite different effects on the same compound, for instance. Again, a given dye may have very different effects on related crystals.

Royer's work on regular intergrowths, which was based on similarities between reticular nets (epitaxis), appeared at about the same time as Bunn's, in 1934. Royer [15] considered that three basic conditions had to be complied with before epitactic intergrowths occurred, namely:

1) the elementary parallelograms of the substances must be almost the same in shape and size;

2) the ions of one substance that replace the ions of the other must have the same polarity;

3) the two lattices must have the same bond type.

It was found possible to explain many cases of habit change caused by traces of impurities. It was found that the impurities could occur as very small oriented crystals, or as two-dimensional precursors thereof.

Royer's theories were confirmed by many experimental facts, and served to explain how some impurities were taken up by growing crystals. Epitactic growths are of more interest to us, however; they were observed later, and relate to substances with differing bond types. Pinsker [16] showed that gold, silver, copper and other metals could form epitactic growths on ionic crystals such as rock salt, fluorite and calcite. His data showed that the surface layers were deformed, often even being twinned.

Dankov [17] observed a more complex relation for crystals with differing bond types; he concluded that the metal atoms lie in potential wells on the faces of the ionic crystal.

Yanulov [18] considered the quantitative aspects of these overgrowths, and found that the dimensional correspondence required by theory was sometimes lacking when the bond types differed. In the rock salt-silver system, for instance, the relative difference in the $\{100\}$ dimensions was 38%, instead of the limit of 15%. Thomson [19] also showed that close correspondence between structures was not always a guarantee of epitactic overgrowth.

Monier's work of 1953-1954 [20, 21] relates to epitaxis in organic crystals of various natures grown from the vapor onto inorganic ionic or homopolar crystals. Naphthalene, anthracene, hexamethylene and tetramine were used to show that crystals of molecular type could show epitaxis if the molecules had no permanent dipole moments, or were unable to form hydrogen bonds to the ionic crystal. If one type of face was especially well developed in the inorganic crystal, then the organic deposited preferentially on that type. If several types were well developed, then all faces behaved in the same way. The most frequent orientation on any given crystal was usually that which gave the best agreement between the parameters, but there were many cases where this rule was not obeyed.

Neuhaus [22, 23] points out that a recent trend in epitaxis studies is to emphasize the energy and kinetic aspects of the process. The type and degree of orientation of the deposit are related to the temperature and hardness of the underlying crystal. Hard materials, if appropriately heat-treated, can orient materials in the same way as soft ones. Thus, feldspar will take oriented KBr crystals at 350°C. Lately, Royer [24] has revised his earlier views, and states that his three conditions need to be reconsidered.

Royer has used new examples of epitaxis to reveal several types of secondary bond important to epitaxis. The phenomena found with overgrowths are more complicated than was at first expected, because the bond strengths and types occur in a great variety of combinations.

Modern ideas on epitaxis assign only secondary importance to similarity between elementary parallelograms.

2. Studies on Crystal Genesis

Crystal genesis work at the end of the 19th century touched on organic impurities in inorganic crystals. The work was directed to finding relations between the composition of the medium and the external form of the crystal. The effects of soluble organic compounds on the habit were examined. The causes of the effects, and the chemical processes in the solutions, were discussed.

Urea was the first compound found to have a major effect on habit. Raumé de l'Isle detected the effect with sodium chloride. Orlov [25], Zemyatchenskii [26, 27] and Spangenberg [28] studied the effects on NaCl systematically. Orlov produced crystals of NaCl, $CH_4N_2O \cdot H_2O$, and observed the octahedra of NaCl that had been reported previously. He supposed that this compound formed because the NH_2 groups in the urea could join with the NaCl to give a double salt. He explained the effect of urea in terms of the changes in composition in the medium caused by such compounds.

Zemyatchenskii's views were similar; he assigned the cause of the habit changes to unstable compounds formed in the solution between the organic and inorganic components. Spangenberg concluded that unstable compounds of double or complex salt type could occur on the faces as well as in the medium. These compounds hindered the growth, changed the relation between face growth-rates, and so altered the form of the crystals. Spangenberg explained the effect of urea on sodium chloride in terms of complex ions $[NaCO(NH_2)_2] \cdot H_2O$ formed on the {111} faces of NaCl. Kuznetsov [29] examined the effects of urea and of other compounds on NH_4Cl, and concluded that the habit was affected only if the organic and inorganic components could react with one another.

In the earlier stages the effects of impurities were held to be related only to the composition and properties of the solution. Orlov and Zamyatchenskii related the habit changes to dissolved compounds, for instance. Retgers [30] found that the impurity affected the surface tension between the solution and the crystal faces.

The reason for these views was that the crystal takes up only minute traces of the organic compound, traces so small as to be undetectable by chemical or x-ray methods. It subsequently became clear that the effects were in fact caused by the minute traces of incorporated impurity. One of the best examples of this is the effect of pectic acid on ammonium chloride, which was first observed by Erlich [31]. The acid causes the habit to change from fibrous dendritic to compact, transparent and elongated. The pectic acid in the crystals was detected indirectly, from repeated crystallization.

We must mention also Sheftal's views. Sheftal' concluded from his work on the effects of the solution on habit (for sucrose, etc.) that any change in habit was related to the part of the solution incorporated in the crystal (solvent, impurity, etc.) [31a].

3. The Main Trends in Studies on Structure and Formation Mechanism for Anomalous Mixed Systems

Some of the work on the structures and formation mechanisms of anomalous mixed systems is of interest. X-ray studies and chemical crystallography have directed much attention to anomalous mixed systems, which are not isomorphous in the original sense. Many examples are known of two unrelated compounds that crystallize together (including ones of inorganic-organic type), and theoretical concepts have been proposed.

Spangenberg and Neuhaus [32] relate the colors of many minerals to the power that substances have to dissolve in one another to give solid solutions or anomalous mixed crystals. Neuhaus [33] considers that anomalous mixed crystals may occur when organic dyes are taken up by inorganic crystals, if the two components have similar reticular structures. Neuhaus follows Gaubert and Bunn in recognizing that the dye may be present in crystalline form.

Lemmlein [34] has studied sector structures in crystals, and considers that morphologically anomalous mixed crystals differ from true isomorphous formations in that the second component is concentrated preferentially on certain faces and in the corresponding growth pyramids. The crystals show sector structures of hour-glass type. Lemmlein considers such crystals anomalously isomorphous. The dye may be present to different extents in the various growth pyramids of such crystals.

Neuhaus [35] has studied barium and lead nitrate crystals containing methylene blue. He has compared the lattices of the two components, and concludes that the structures are very similar, if the nitrate (and not the chloride) of the dye is considered. He argues that space-lattice considerations imply that anomalous crystals should be considered as mixed crystals of variable composition, if the lattice and cell volumes are very similar.

He relates the origin of these mixed crystals to the tendency for the two components to form complexes in the solution. The mixed crystals are assumed to be produced as follows. The growth is rhythmic, and the impurity concentration at the growing surfaces varies periodically. At first, only the main component is deposited in the solid, because the solution is not saturated in the minor component. The concentration of the latter rises as crystallization proceeds. The tendency to complexing causes diffusion at the surfaces to slow down; the surface layers become richer in impurity. If the growth rate is high enough, it becomes possible for nuclei to form at isolated points where the supersaturation is high. Volmer [36] has shown that the energy needed to produce such nuclei is quite small. The two structures are similar, so the two substances crystallize together, and the supersaturation is reduced.

Neuhaus' data on strontium nitrate (which takes up methylene blue little or not at all) are quoted to demonstrate the role of complexing. This effect is related to the tendency of strontium not to form complexes readily. This example is not a very good one for the purpose, however, because we have been able to make strontium nitrate crystals containing methylene blue. Not only so; the dye was taken up at all concentrations, which is not so with lead and barium nitrates.

Grimm and Goldschmidt's work on anomalous mixed crystals, in which isomorphism is divided into sorts one, two and three [37], is of great importance. Grimm relates the tendency of inorganic ionic compounds to crystallize together to the properties of the constituent ions. A major factor here is that the radii of the corresponding ions in the two components should be similar; mixed crystals formed between unrelated components (isomorphism types two and three) occur because isomorphous compensation is operative. The mixed crystals formed by $BaSO_4$ and $KMnO_4$ are typical examples of Grimm's type-two isomorphism.

Stranski [38] assumes another mechanism for such processes. Crystals that are isomorphous in Grimm's sense can occur only under certain conditions. For example, a crystal composed of univalent ions can grow on an isomorphous one composed of divalent ions when the solution is unsaturated with respect to the first component. Large supersaturations in the second component are needed to cause the reverse process, however .

Balarev [39] doubts whether such systems can be said to crystallize together at all; he holds that in the $BaSO_4-KMnO_4$ system the potassium permanganate is adsorbed on internal surfaces (pores) in the barium sulfate.

The work of Khlopin's school is of special importance; it has been concerned with anomalous mixed systems that have Grimm's new types of isomorphism, and with systems that are (according to Khlopin) structurally related to them, namely, inorganic salt plus organic dye. The main systems used were 1) $BaSO_4-KMnO_4$ and 2) barium (or lead) nitrate with methylene blue.

Khlopin sought to establish whether such mixed systems are formed because the components crystallize together, or whether adsorption or some other effect was responsible. He studied the structure of the system in terms of the distribution of the minor component between the crystalline and liquid phases.

Khlopin first of all found [40] that a trace isomorphous component in such a system was distributed in accordance with Nernst's law. The distribution constant did not vary, because the trace component had no practical effect on the compositions of the phases; the concentration in the crystal was proportional to that in the solution. These mixed systems are formed by isomorphous syncrystallization, and so the distribution should follow Nernst's law.

Nikitin [41] found that in the $BaSO_4-KMnO_4$ system the second component was absorbed by the first only when the concentration of the second was above some limit, but that Nernst's law was then obeyed. Khlopin and Nikitin explain this lower bound to the miscibility in terms of complete replacement of parts of one lattice by the other, rather than replacement ion-by-ion. The result is an extremely finely divided mixture of the two substances, which appears homogeneous.

Khlopin and Tolstaya [42] and later É. M. Ioffe and Nikitin [43, 44], examined the barium (lead) nitrate-methylene blue system, in which lower bounds to the miscibility had also been found. These bounds were found to be stable against third components; this, with the reverse distribution,* indicated that the dye was not adsorbed, but was actually crystallizing with the nitrate.

* By this is meant the uptake of barium nitrate into growing methylene blue crystals.

Further studies on the distribution in such systems showed that there were deviations from Nernst's law, and so it was supposed that the structures of these crystals differed considerably from those of isomorphous ones. Each small crystal in such a system has a regular lattice, in isolated areas of which the ions of the trace component are distributed with statistical regularity [44]. These areas are large relative to those envisaged in Grimm's theory of substitution in mixed crystals.

Khlopin's school has thus shown that a dye can enter a nonisomorphous crystal from an unsaturated solution.

The structures and formation mechanisms of systems in which an inorganic salt contains an organic dye as a trace component have as yet not been established.

France's Researches

It remains to consider France's work, in which ideas of adsorption are used to explain how dyes are taken up by inorganic crystals. The work is of interest because the type of adsorption is studied in relation to the ionic structure of the crystal faces.

When x-ray methods became available, adsorption effects (especially selective adsorption) began to be studied in relation to the structures of the adsorbing surfaces. Marc [45, 46] studied the effects of dyes on the crystallization kinetics of several substances, and found that the dyes affected the face growth-rates differently.

Niggli [47] argues that the growth rate of a face, i. e., the rate of uptake of material, is determined by the unsatisfied residual valencies on the surface, whose number depends on the reticular density of the face. Valeton's theory [48] is that the electrostatic charges of the ions in the surface layer attract ions of opposite sign from the solution. The force field of a face with an array of ions of opposite signs should thus be weaker than the field of a face with ions all of one sign.

Langmuir's theory of surface-active materials shows that the activity of the surface of a solid depends on the spatial array of the atoms or ions on that surface.

France [49, 50] used these theories in his work, and concluded that, in the main, the adsorption effects found with dyes agreed with these theories. His data on relative face growth-rates for alum, lead nitrate and other substances in the presence of dyes led him to the following conclusions.

1. The adsorption is usually strongest on faces populated by ions of one sign only. If all the faces are of this type, the adsorption is strongest on those that have the strongest electric fields. The field strengths are determined by the distances between the ions.

2. The adsorption depends on the number and positions of the polar groups in the dye.

3. The amount of dye adsorbed is much less than that needed to give a continuous monomolecular film.

France observes that some exceptions to these rules occur. Some alums, for example, although they have the same structure and very similar ionic spacings, adsorb a given dye on different faces. Buckley has found (for K_2SO_4) that the dye may be adsorbed by faces having ions of both signs when other faces have ions of one sign only [51]; Frondel [52] has found the same for sodium fluoride. In the latter case the neutral {100} faces take up the dye. The reverse pattern also occurs; a given alum reacts differently to closely related dyes.

France has crystallized lead nitrate in the presence of methylene blue of bismarck brown, and concludes that here the adsorption has causes more complex than the mere ion arrays on the surfaces. He found that methylene blue gave dark-colored cubes, whereas bismarck brown gave tetrahedra that were just as dark in color. These facts by themselves show that cube and tetrahedron faces can both adsorb large amounts of dye. France also takes a view similar to Buckley's (p. 8) on the way the dye is held by the crystal. Both of these workers suppose that a dye molecule is held to the surface by one end only, with the other end free.

France's general conclusion is that there is no simple rule with which to predict which organic compounds will be adsorbed by any given crystal.

To sum up, we may say that there are two opposed views on the interactions between inorganic crystals and salt-like dyes present in trace amounts.

The first view (Gaubert, Buckley, France) is that these dyes can be present in the crystals only as single molecules, which, according to Buckley, are oriented; this explains the pleochroism often seen in such crystals.

Such systems must be considered as solid solutions, because the dye is present in molecular form. However, it has still not been decided whether an organic compound can form a true solid solution in an inorganic crystal, although lately the topic has not received the attention it should have.

The second view (Bunn, Neuhaus, Khlopin) is that the trace component may be present in crystalline form in the host crystal. A notable point here is that there is no major disagreement between the various workers on the structures and formation mechanisms of anomalous mixed crystals formed from two components present in comparable amounts. These crystals are anomalously isomorphous, and obey the laws of epitactic overgrowth, or of isomorphous substitution. The disagreements over crystals formed from a major and a trace component are many, and there are many ideas and hypotheses on the structures and formation mechanisms.

The problem is one of interest to crystallographers and to physical chemists. It is far from clear how the problem is to be solved, because not many ways have yet been found to grow such crystals under these conditions.

Khlopin's work is of interest in this context; he holds that the dye must be present as submicroscopic crystals in the main crystal. These crystals have not been detected by direct x-ray methods, because they are so small. Tatarinova's electron-diffraction studies [42] have shown that methylene blue is present in crystalline form in barium nitrate crystals, but the conditions were such as to prove that the state was crystalline only for high concentrations of dye in the solution.

Vedeneeva's spectrophotometric studies of the adsorbed state for dyes [1, 2] give us a fresh approach to the problem. It would be desirable to study the states of such dyes in relation to the types of bond formed between the dye molecules (cations) and the surfaces of the host crystal.

In 1952 Vedeneeva and Slavnova [53] attempted to apply this method to methylene blue in barium and lead nitrate crystals. Vedeneeva's absorption spectra of methylene blue as crystals [53] and in the adsorbed state were used in this work.

We found that the state of the dye could differ even within the crystals of a single isomorphous series. In $Pb(NO_3)_2$ the dye was present in molecular form, while in $Ba(NO_3)_2$ it was present in crystalline form.

Detailed studies showed that the crystals were determined as to orientation in any growth pyramid by the structure of the corresponding face, and by the dye concentration in the solution. If the concentration was low, the crystals were so oriented that the dye molecules usually had their planes parallel to the faces (or nearly so). At high concentrations the molecular planes (and the long axes of the crystals) were usually normal to the faces. The density of the dye crystals on the surfaces increases with the dye concentration in the solution [54-56].

The cases where the dye was present in molecular form may perhaps be ascribed to the formation of a special type of solid solution. It is necessary for the faces to bind the single dye molecules strongly if structural elements of the host crystal are to be replaced by dye molecules.

In other words, if the structural and energy aspects are satisfied, the dye concentration must be sufficient to cause the dye to enter the crystal, but small enough for any aggregation on the part of the dye molecules on the surfaces to be negligible.

The concentrations in such systems are so low ($4.0 \cdot 10^{-2}\%$ in the case of lead nitrate) that the usual physicochemical methods are not applicable. Therefore our criteria for the formation of solid solutions in such cases may be of some interest.

In this connection we must note Whetstone's work [57, 58], in which the configurations of some trimethylene dyes were compared with the structures of certain crystals (ammonium nitrate and sulfate, among others); Whetstone believes that here the dyes form solid solutions in the crystals. A necessary condition is that the polar groups in the molecule should correspond to the structure units that are replaced in the crystal.

LITERATURE CITED

[1] P. A. Rebinder, Surface Effects, Adsorption, and the Properties of Adsorbed Films [in Russian] (1932).

[2] N. E. Vedeneeva, Zhur. Fiz. Khim. 21, 883-891 (1948); Kolloid Zh. 12, 1, 17-24 (1950); Doklady Akad. Nauk SSSR 98, 4, 585-8 (1954).

[3] H. Senarmont, Ann. chim. 41, 319-336 (1854).

[4] P. Gaubert, Bull. Soc. Franc. Minerol. et cristallogr. 17, 121-123 (1894).

[5] P. Gaubert, Compt. rend. 155, 649-651 (1912).

[6] P. Gaubert, Compt. rend. 180, 378 (1925).

[7] P. Gaubert, Compt. rend. 194, 109 (1932).

[8] P. Gaubert, Publ. Soc. Chim. Phys., Paris 1 33 (1911).

[9] C. Bunn, Proc. Roy. Soc. A 141, 567-593 (1933).

[10] C. Bunn and Emmett, Disc. Faraday Soc. 5, 119-132 (1949).

[11] H. E. Buckley, Crystal Growth (J. Wiley, N. Y., 1951).

[12] H. Buckley and Cocker, Z. Krist. 85, 58-75 (1933).

[13] H. Buckley, Mem. Proc. Manchester Lit. Phil. Soc. 83, 31-62 (1939).

[14] H. Buckley, Z. Krist. 76, 147 (1930).

[15] Royer, Bull. Soc. Franc. mineral. et cristallogr. 51, 7-159 (1928).

[16] Z. G. Pinsker, Electron Diffraction [in Russian] (Izd. AN SSSR, 1949), pp. 211-238.

[17] P. D. Dankov, Zhur. Fiz. Khim. 20, 853-867 (1946).

[18] N. P. Yanulov, Epitactic Growth in Crystals (Author's abstract of dissertation as candidate in geological-mineralogical sciences) [in Russian] (Leningrad, 1951).

[19] G. Thomson, Proc. Phys. Soc. 61, 403 (1948).

[20] J. Monier, Compt. rend. Acad. Sci. 236, 21, 2089-2091 (1953).

[21] J. Monier and H. Raymond, Bull. Soc. franc. mineral. et cristallogr. 77, 1029-1048 (1954).

[22] A. Neuhaus, Z. angew. Chem. 6, 147-148 (1950).

[23] A. Neuhaus, Angew. Chem. 6, 158-162 (1952).

[24] L. Royer, Bull. Soc. franc. Mineral. et cristallogr. 71, 1004-1029 (1954).

[25] P. P. Orlov, Zap. Mosk. Obshch. Ispyt. Prirod. 4, 1-60 (1896).

[26] P. A. Zamyatchenskii, Zap. Imp. Akad. Nauk, series 8, fiz.-mat.-otedel. 24, 8, 1-36 (1909).

[27] P. A. Zamyatchenskii, Zap. Imp. Akad. Nauk, series 8, fiz.-mat.-otdel. 30, 3, 1-19 (1911).

[28] K. Spangenberg, Z. Krist. 59, 383-405 (1924).

[29] V. D. Kuznetsov, Crystals and Crystallization [in Russian] (Gostekhizdat, 1953), pp. 234-5.

[30] J. W. Retgers, Z. phys. Chem. 9, 267-311 (1892).

[31] F. Ehrlich, Z. anorg. u. allgem. Chem. 203, 26-38 (1932).

[31a] N. N. Sheftal', Growth of Crystals, I [in Russian] (Izd. AN SSSR, 1957), pp. 5-31.*

[32] K. Spangenberg and A. Neuhaus, Chemie Erde 5, 437-538 (1930).

[33] A. Neuhaus, Chemie Erde 5, 529-555 (1930).

[34] G. G. Lemmlein, Sector Structures in Crystals [in Russian] (Izd. AN SSSR, 1948), pp. 8-13.

[35] A. Neuhaus, Z. Krist. 103, 5, 297-326 (1941).

---

* See C.B. translation, page 5.

[36] M. Volmer, Kinetik d. Phasenbildung (Dresden, 1939).

[37] V. M. Goldschmidt, Chemical Crystallography [Russian translation] (ONTI, Leningrad, 1937), pp. 16-20 and 54-7.

[38] J. Stranski, Z. phys. Chem. A 142, 453-466 (1929).

[39] D. Balarew, Z. anorg. u. allgem. Chem. 174, 295-347 (1928).

[40] V. G. Khlopin, Trudy Gos. Rad. Inst. 4 34-84 (1938).

[41] B. A. Nikitin, Trudy Gos. Rad. Inst. (1930).

[42] V. G. Khlopin and M. A. Tolstaya, Zhur. Fiz. Khim. 14, 7, 941-952 (1940).

[43] É. M. Ioffe and B. A. Nikitin, Izv. AN SSSR, Otdel. Khim. Nauk 1, 15-22 (1943).

[44] É. M. Ioffe and B. A. Nikitin, Izv. AN SSSR, Otdel. Khim. Nauk 3, 191-7 (1943).

[45] R. Marc and W. Wenk, Z. phys. Chem. 68, 104-114 (1909).

[46] R. Marc, Z. Elektrochem. 20, 515-521 (1914).

[47] P. Niggli, Z. anorg. Chem. 110, 55-80 (1920).

[48] J. Valeton, Physik Z. 21, 606 (1920); 59, 135, 335 (1924).

[49] W. France, Reprinted from Colloid Symposium Annual 60-87 (1930).

[50] W. France, Colloid Chemistry 5, 443-457 (1944).

[51] H. Buckley, Z. Krist. A 88, 248-255 (1934).

[52] C. Frondel, Am. Mineralogist 25, 91 (1940).

[53] N. E. Vedeneeva and E. N. Slavnova, Trudy Inst. Krist. 7, 135-158 (1952).

[54] E. N. Slavnova, Interaction of Traces of Organic Dyes with the Faces of a Growing Crystal (Author's abstract of dissertation as candidate in technical sciences) [in Russian] (1956).

[55] E. N. Slavnova, Doklady Akad. Nauk SSSR 106, 6, 1007-1010 (1956).

[56] E. N. Slavnova, Growth of Crystals, I [in Russian] (Izd. AN SSSR, 1957).*

[57] J. Whetstone, Disc. Faraday Soc. 16, 132-140 (1954).

[58] J. Whetstone, Trans. Faraday Soc. 51, 8, 1142-1153 (1955); J. Chem. Soc. 4841-4847 (1956).

*See C.B. translation, page 117.

THE THEORY OF THE CRYSTALLIZATION OF STEEL*

P. S. Vadilo

Crystallization problems arise in crystallography and in metallurgy. In metallurgy we encounter opaque polycrystalline substances, which are much more difficult to examine for structure than are transparent substances or single crystals. The interchange of ideas between crystallography and metallurgy is not good.

The data on the crystallization of castings are scattered over many journals, and so it is difficult for crystallographers to become acquainted with the ideas prevalent in that field. The recent Russian translation (from the Czech) of Hvorinov's book "Crystallization and Inhomogeneity in Steel" [1] is of some considerable assistance in this respect. This book is spoken of highly by Soviet metallurgists and metallographers, and contains a survey of current ideas on crystallization of metals, as well as some novel ideas of the author's own on steel.

I have used this book in order to discuss crystal division (multiplication), which is an effect well known to crystallographers, but which is neglected by metallurgists. This neglect is unfortunate. Tamman's view (that the number of crystallites in a casting is determined mainly by the processes that give spontaneous nucleation) long prevailed in metallurgy. This view has been replaced by the more correct one that nucleation occurs mainly on impurity particles that are isomorphous with or epitactic to the main component, and that are present before crystallization starts. The crystal division processes are closely related to movement in the melt, but have been neglected by metallurgists. Nevertheless, these processes play a large part in nucleation. This paper deals with these topics in relation to steel. It has been compiled with assistance from Hvorinov's book, from published data, and from my own experiments.

Hvorinov describes methods of crystallizing steel that improve the quality of the castings, and propounds a new theory of crystallization. This theory derives from studies made at the Lenin Works, and at the Special Steels Works, in Czechoslovakia, and also from laboratory researches. Results are given on the inner surfaces of steel castings from which the central liquid has been withdrawn during crystallization. Photographs of the surfaces are given, as well as ones of fracture surfaces and of etched polished surfaces.

Hvorinov considers that dendrites form at the edge (where the supercooling is large) when the steel is poured into the mold. Dendrites do not arise spontaneously in central areas. They occur only as branches of edge dendrites, which may have been broken or melted off by random or forced movements in the liquid steel. These movements carry the pieces into the center, where they grow as independent (usually equiaxial) dendrites. These dendrites become the primary grains.

Liquation occurs and is caused mainly by two factors, namely, on account of major mechanical disturbances in the melt when it is full of dendrites, and on account of hydrogen bubbles which escape and leave behind areas that are filled by liquid enriched in impurities.

Here we mean by disturbances those responsible for cracks and the like in the semisolid material, which disturbances may break the dendrites away from the walls of the mold. They may be produced by rocking or twisting the mold violently, or by pressing on the casting. If the motion is of appropriate frequency and amplitude (e. g., gentle rocking in a horizontal plane, or ultrasonic vibration), the dendrites are broken away from the walls and are carried to the center, but no liquation occurs. The structure and quality of the casting are improved.

---

* For discussion.

It can hardly be doubted that Hvorinov's methods of improving the quality of castings should be given serioud attention. There are, however, other methods of improvement, which may turn out to be more effective; these we deal with below.

It may be agreed that Hvorinov's theory is better than earlier theories. There are some parts of it, however, with which it is difficult or impossible to agree.

Hvorinov's list of references relates only to work on metallurgy. There is, however, nothing very exceptional about steel. It is wrong to neglect the work of crystallographers and mineralogists, and to do so weakens Hvorinov's arguments, since his theory does not correspond exactly to the true state of affairs.

It is quite possible that convection and artificial stirring may move dendrites from the walls to central areas, where they grow as independent equiaxial dendrites. But it is hard to agree with the view that the central dendrites arise only in this way. Equiaxial dendrites are usually much commoner in the central areas than are needle-like or columnar crystals at the edges. In some cases the edges consist of large columnar crystals, and the central areas, of large monocrystal grains. In such cases dendrites are entirely absent.

To support the doubts expressed here we may mention data derived from crystallization from solutions, for which the regularities that are found do not in principle differ from the regularities found for steel melts. We give also some data on melts of organic compounds and of fusible metals.

1. The silver nitrate and potassium bromide solutions used to make photographic emulsions are stirred very vigorously for long periods. This is done to ensure that the grain size is small. The silver bromide grains are not dendrites and so we cannot say that the grain size is reduced because dendrites are broken up.

2. We have used slightly and highly supersaturated solutions of ordinary alum (5 g and 500 g excess solid per liter, respectively), and have found that they do not crystallize when they are stirred or shaken vigorously even if growing alum crystals are present. In both cases, however, we found enormous numbers of very small crystals if the growing crystal struck the walls, the stirrer or another growing crystal. These blows detached minute particles which then grew as independent crystals.

It is clear from this that the numerous small crystals are formed from minute fragments broken from normal crystals, and not by fracture and transfer from dendrites. The same effect may, of course, occur with dendrites , but Hvorinov takes no account of this.

3. We allowed 1 kg of molten paraffin wax to crystallize without disturbance; the solid had a shrinkage hole that covered the entire free surface. The hole extended down to the bottom of the vessel. If, however, the molten paraffin was stirred with a rod heated to a temperature slightly above the melting point of paraffin wax, there was no shrinkage hole. No cracks of holes in the solid could be detected by eye. The cause is, of course, that the stirring made the temperature much more uniform. The crystals collided repeatedly, and grew in all parts at about the same rate.

4. 1 kg of lead was cast in an iron mold; dendrites up to 3 cm in diameter were formed. There were also shrinkage cracks, hollows and holes. A similar casting in which the molten metal had been stirred showed no dendrites, but there were small octahedral crystals (less than a millimeter across) and monocrystalline grains. There was no shrinkage pit, nor any cracks or holes. The cause is simply that the stirring evened out the temperature and caused the crystals to spread to all parts. The melt became a gradually thickening suspension of small crystals. This gave rise to the fine-grained hole-free casting.

Thus vigorous stirring aids crystal growth and evens out the temperature, and so gives fine-grained castings; cracks, columnar crystals, shrinkage pits, etc. are eliminated.

The mechanism of crystallization , and the relation of this process to supersaturation (supercooling) and to stirring can be deduced from the following simple experiments.

5. One liter of ammonium sulfate heptahydrate solution (saturated at 40°C) was heated to boiling in a closed vessel. The solution did not crystallize on cooling to room temperature, in spite of vigorous stirring and shaking. The solution was again heated to boiling, and a small glass tube was added containing crystals of the salt in contact with a solution saturated at room temperature. One end of this tube was sealed; the other was drawn out into a capillary. The whole was sealed and was allowed to stand. The crystals in the tube at first began to dissolve, but did not dissolve completely. One of the residual crystals then grew out through the capillary and became a

dendrite with several branches. The concentration flows caused by this dendrite broke off the thinnest branches and carried them up into the body of the solution. As soon as 3 or 4 such branches had been carried up the entire mass of the solution was filled with thousands of needle-like dendrites, which formed a mass in which the solution, no longer supersaturated, was trapped.

These numerous crystals formed from a few branches because the latter collided with one another and with the walls of the vessel. The blows knocked minute particles off the dendrites, which particles then became new crystals.

Here the crystals multiplied because the concentration currents, which were vigorous, moved the crystals about.

6. A supersaturated solution of the same salt, with a crystal growing in it, was stirred vigorously; then, no matter what the supersaturation, there arose numerous small isometric crystals (less than a millimeter across). Here the stirrer, the walls, and the crystals themselves, caused the knocks. These knocks were harder and more frequent than those produced by the concentration currents, so the crystals multiplied very much more rapidly, and the solution soon become only saturated; no dendrites could form. Dendrites occur only with highly supercooled or supersaturated media.

We now return to Hvorinov's theory; the above evidence shows that Hvorinov's idea that dendrites or crystals form only in areas near the walls is unproven.

Steel can be supercooled by more than 200°C without giving crystals or dendrites [2]. The number of branches in a dendrite increases with the cooling rate, whereas the number of dendrites does not [3]. The dendrites or crystals do not arise in the way assumed in Tamman's theory. Nuclei are required, which must be small particles of the substance itself, or of isomorphous or epitactic substances [4]. These nuclei may be present at the walls, or may fall into the melt from the air. Most of the nuclei are produced by knocks between the few original crystals.

This very important source of nuclei is entirely neglected in Tamman's and Hvorinov's theories.

The melt is very much supercooled at the edges when it is poured into a mold. The nuclei present at the walls give crystals, whose number is increased by the above multiplication process. Of the crystals produced initially at the walls, the only ones to continue growing are those whose vertices point towards the center and which are in contact with the main mass of melt, but not with nearby crystals. These crystals grow as needles if the supercooling is large, and sometimes have dendritic branches at their free ends; they become columnar if the supercooling is slight.

The concentration currents [6] break off the branches and spread them about the melt. The branches collide with one another or with the walls, and give off small particles. These, and the branches, then become independent equiaxial dendrites. The crystals multiply more rapidly if the liquid part of the casting is disturbed by shaking the mold, or with ultrasonics. The grain will become finer. The crystals that grow if the supercooling is slight are those initiated by particles present on the walls or in the melt, or entering from the air. Few crystals arise in the central areas, where large monocrystalline grains are found; the entire casting may be made up from a few columnar crystals.

Cracks, holes and even shrinkage pits are bound to occur if the metal crystallizes from a crust of needle-like or columnar crystals. The same effect occurs in the continuous casting process. These types of crystal may penetrate to the central areas from the walls even while the steel is being poured, and so may give rise to a trapped body of liquid. This liquid is insufficient to fill the volume available when it cools down and crystallizes. The result is shrinkage pits and pores between the branches of the dendrites.

It would be possible to make castings free from defects (and not merely with less defects, as in continuous pouring, or in Hvorinov's method) by removing the region of columnar crystals by methods which accelerate multiplication and temperature equalization.

These effects cannot be produced by rotating or rocking the mold slowly, or with ultrasonics, as in Hvorinov's method. The temperature is not thereby evened out, the columnar crystals still form their zone, and numerous small crystals do not arise. Vigorous stirring is needed, because it evens out the temperature, distributes the crystals uniformly, and encourages multiplication. A crust of solid metal cannot form, nor can columnar crystals. The melt becomes a gradually thickening suspension of very small crystals (not dendrites), which sets to a fine-grained casting free from pits, cracks and pores. It may also be that the stirring will prevent the impurities from causing liquation, because they are evenly distributed between the numerous small crystals; the quality of the casting may be improved.

Hvorinov suggests that the mold should be turned in a horizontal plane in order to encourage fusion and breakage of the dendrites, and to improve transfer of broken branches to the center. It may be that the fusion and breakage would be improved by the turning, but the pieces would be thrown towards the walls, which would mean that pores and shrinkage pits would occur in the center, as is the fact found with Hvorinov's castings.

A motion of planetary type would throw the fragments in part towards the center, and might assist in evening out the temperature. Vigorous use of the stirrer should give better results.

Some of Hvorinov's arguments in support of his view about the origin of the primary grains from few or many dendrites are, to say the least, unconvincing.

Hvorinov's photographs of etched primary grains show branches of dendrites that sometimes are clearly different in orientation. Hvorinov considers that these branches belong to different dendrites. It is hard to see why this should be so. The feather and fern patterns seen on frosted windowpanes consist of branches that have geometrically distinct orientations. They are, nonetheless, the branches of a single dendrite, because the branches of one fern all reflect the sun's rays in the same direction, whereas those of another fern reflect in another direction. Hvorinov's photographs of etched grains also show fern-like dendrites. A single primary grain is formed from one such dendrite; it preserves the latter's crystallographic orientation. The dendrite may trap isolated needle-like crystals and branches of other dendrites, which may also be seen in the etched primary grain.

LITERATURE CITED

[1] N. I. Hvorinov, Crystallization and Inhomogeneity in Steel [Russian translation] (Moscow, 1958).

[2] I. S. Gaev, A Metallographic Atlas [in Russian] (Moscow, 1941).

[3] P. S. Vadilo, Zap. Kishinev. Univ. 1, 1 (1949).

[4] P. S. Vadilo, Zap. Kishinev. Univ. 3, 1 (1951).

[5] D. S. Kamentskaya, Growth of Crystals, I [in Russian] (Izd. AN SSSR, 1957).

[6] L. N. Matusevich, Growth of Crystals, I [in Russian] (Izd. AN SSSR, 1957).